Lecture Notes in Computer Science 10819

Commenced Publication in 1973
Founding and Former Series Editors:
Gerhard Goos, Juris Hartmanis, and Jan van Leeuwen

Editorial Board

More information about this series at http://www.springer.com/series/7409

Miguel R. Luaces · Farid Karimipour (Eds.)

Web and Wireless Geographical Information Systems

16th International Symposium, W2GIS 2018
A Coruña, Spain, May 21–22, 2018
Proceedings

 Springer

Editors
Miguel R. Luaces
Universidade da Coruña
A Coruña
Spain

Farid Karimipour ⓘ
University of Tehran
Tehran
Iran

ISSN 0302-9743　　　　　　ISSN 1611-3349　(electronic)
Lecture Notes in Computer Science
ISBN 978-3-319-90052-0　　　　ISBN 978-3-319-90053-7　(eBook)
https://doi.org/10.1007/978-3-319-90053-7

Library of Congress Control Number: 2018939623

LNCS Sublibrary: SL3 – Information Systems and Applications, incl. Internet/Web, and HCI

Printed on acid-free paper

This Springer imprint is published by the registered company Springer International Publishing AG
part of Springer Nature
The registered company address is: Gewerbestrasse 11, 6330 Cham, Switzerland

Preface

Recent developments in wireless Internet technologies have generated an increasing interest in the diffusion and processing of a large volume of geo-spatial information. Spatially enabled wireless and Internet devices also offer new ways of accessing and analyzing this geo-referenced information in both realworld and virtual spaces. Consequently, new challenges and opportunities have appeared in the GIS research community.

These proceedings contain the papers selected for presentation at the 16th edition of the International Symposium on Web and Wireless Geographical Information Systems (W2GIS) held in A Coruña, Spain, in May 2018 and hosted by the University of A Coruña. This symposium is intended to provide an up-to-date review of advances in both theoretical and technical development of Web and wireless GIS. It was the 16th in a series of successful events beginning with Kyoto 2001, and mainly alternating locations annually between East Asia and Europe. It provides an international forum for discussing advances in theoretical, technical, and practical issues in the field of wireless and Internet technologies suited for the dissemination, usage, and processing of geo-referenced data.

W2GIS is organised as a full two-day symposium and is recognized as a leading forum for dissemination and discussion on the latest research and development achievements in the Web GIS and wireless domains. The submission process was successful this year, attracting papers from almost all over the world. This demonstrates not only the growing importance of this field for researchers but also the growing impact these developments have in the daily lives of all citizens.

Each paper received at least three reviews and was ranked accordingly. The accepted papers are all of excellent quality and cover topics that range from Web technologies and techniques, paths and navigation, location-based service, Web visualization, and novel applications.

We wish to thank the authors who contributed to this symposium, for the high quality of their papers and presentations, and the Springer LNCS team, for their support. We would also like to thank the Program Committee for the quality and timeliness of their evaluations. Many thanks to the Steering Committee for providing continuous advice and recommendations. Finally, we appreciate the local Organizing Committee for their continuous support, and Kimia Amouzandeh, from the University of Tehran, for her contributions in the process of organizing the conference proceedings.

March 2018

Miguel R. Luaces
Farid Karimipour

Organization

Symposium Chairs

M. R. Luaces   Universidade da Coruña, Spain
F. Karimipour   University of Tehran, Iran

Local Chairs

O. Pedreira   Universidade da Coruña, Spain
G. de Bernardo   Universidade da Coruña, Spain
A. Cortiñas   Universidade da Coruña, Spain

Proceedings Organizer

K. Amouzandeh   University of Tehran, Iran

Steering Committee

M. Bertolotto   University College Dublin, Ireland
J. D. Carswell   Dublin Institute of Technology, Ireland
C. Claramunt   Naval Academy Research Institute, France
M. Egenhofer   NCGIA, USA
J. Gensel   University of Grenoble, France
K. J. Li   Pusan National University, South Korea
S. Liang   University of Calgary, Canada
K. Sumiya   Kwansei Gakuin University, Japan
T. Tezuka   Tsukuba University, Japan
C. Vangenot   University of Geneva, Switzerland

Program Committee

Masatoshi Arikawa   The University of Tokyo, Japan
Andrea Ballatore   Birkbeck, University of London, UK
Scott Bell   University of Saskatchewan, Canada
Michela Bertolotto   University College Dublin, Ireland
Alain Bouju   La Rochelle University, France
David Brosset   Naval Academy Research Institute, France
Elena Camossi   Centre for Maritime Research and Experimentation,
                               Italy
James Carswell   Dublin Institute of Technology, Ireland
Christophe Claramunt   Naval Academy Research Institute, France
Maria Luisa Damiani   University of Milan, Italy

Sergio di Martino	University of Naples, Italy
Max Egenhofer	University of Maine, USA
Zhixiang Fang	Wuhan University, China
Filomena Ferrucci	Università di Salerno, Italy
Stefan Funke	University of Stuttgart, Germany
Jérome Gensel	Laboratoire d'Informatique de Grenoble, France
Ralf Hartmut Güting	Fernuniversität Hagen, Germany
Farshad Hakimpour	University of Tehran, Iran
Bo Huang	The Chinese University of Hong Kong, SAR China
Haosheng Huang	University of Zurich, Switzerland
Yoshiharu Ishikawa	Nagoya University, Japan
Farid Karimipour	University of Tehran, Iran
Ki-Joune Li	Pusan National University, South Korea
Songnian Li	Ryerson University, Canada
Xiang Li	East China Normal University, China
Steve Liang	University of Calgary, Canada
Yu Liu	Peking University, China
Miguel R. Luaces	Universidade da Coruña, Spain
Bruno Martins	Universidade de Lisboa, Portugal
Miguel Mata	UPIITA-IPN, Mexico
Sebastien Mustiere	IGN, France
Kostas Patroumpas	National Technical University of Athens, Greece
Dieter Pfoser	George Mason University, USA
Cyril Ray	Ecole Navale, France
Kai-Florian Richter	University of Umea, Sweden
José Ramón Ríos	University of Santiago de Compostela, Spain
Maribel Yasmina Santos	University of Minho, Portugal
Markus Schneider	University of Florida, USA
Diego Seco	University of Concepción, Chile
Sabine Storandt	JMU Würzburg, Germany
Kazutoshi Sumiya	Kwansei University, Japan
Taro Tazuka	University of Tsukuba, Japan
Yannis Theodoridis	University of Piraeus, Greece
Martin Tomko	University of Zurich, Switzerland
Christelle Vangenot	University of Geneva, Switzerland
Marlene Villanova-Ol	Laboratoire d'Informatique de Grenoble, France
Robert Weibel	University of Zurich, Switzerland
Stephan Winter	The University of Melbourne, Australia
Jing Wu	East China University of Technology, China
Javier Zarazaga	University of Zaragoza, Spain
Danielle Ziebelin	Université Grenoble-Alpes, France

Additional Reviewer

Asensio, Angel	University of Zaragoza, Spain

Contents

Improving Sensing Coverage of Probe Vehicles with Probabilistic Routing

Dario Asprone[1], Sergio Di Martino[1(✉)] [iD], and Paola Festa[2]

[1] Department of Electrical and Telecommunications Engineering,
University of Naples Federico II, Naples, Italy
dario.asprone@gmail.com, sergio.dimartino@unina.it
[2] Department of Mathematics and Applications "R. Caccioppoli",
University of Naples Federico II, Naples, Italy
paola.festa@unina.it

Abstract. Modern cars are pervasively equipped with multiple sensors meant to improve in-vehicle quality of life, efficiency and safety. The aggregation on a remote back-end of the information collected from these sensors may give rise to one of the biggest and most pervasive sensor networks around the world, making possible to extract new knowledge, or contextual awareness, in a detail never experienced before. Anyhow, an open issue with probe vehicles is the achievable spatio-temporal sensing coverage, since vehicles are not uniformly distributed over the road network, because drivers mostly select a shortest time path to destination. In this paper we present an evolution of the standard **A*** algorithm, where the route is chosen in a probabilistic way, with the goal to maximize the spatio-temporal coverage of probe vehicles. The proposed algorithm has been empirically evaluated by means of a public dataset of more than 320.000 real taxi trajectories, showing promising performances in terms of achievable sensing coverage.

1 Introduction

Modern cars are pervasively equipped with sensors meant to improve the vehicle efficiency, as well as the quality of life and safety of passengers. Indeed, the deployment of *Advanced Driver Assistance Systems* (ADAS) requires a wide array of sensors, like frontal and backward radars, frontal/surrounding cameras, surrounding ultrasonic sensors, and so on, to get a picture of the environment surrounding the vehicle, like relative position of pedestrians, bikes, and other vehicles. A limited list of the sensors meant to support ADAS is shown in Fig. 1.

Very interestingly, the aggregation on a remote back-end of the information collected from the sensors of a fleet of vehicles may give rise to one of the biggest and most pervasive sensor networks around the world, making possible to develop a collective intelligence, or contextual awareness, in a detail never experienced before [1]. Indeed, while a vehicle drives by a road segment, its sensors constantly scan the environment. The location of each information of interest can be sent by the vehicle to a back-end server via the cellular network [2], where

© Springer International Publishing AG, part of Springer Nature 2018
M. R. Luaces and F. Karimipour (Eds.): W2GIS 2018, LNCS 10819, pp. 1–10, 2018.
https://doi.org/10.1007/978-3-319-90053-7_1

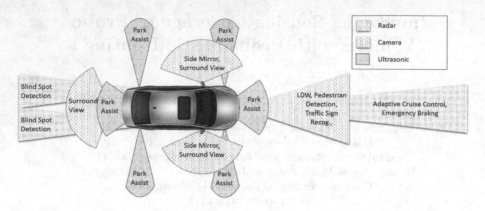

Fig. 1. Some of the vehicle's sensors for ADAS.

it can be aggregated with the data coming from other vehicles, and processed to generate new contextual knowledge. As an example, a significant fraction of modern vehicles have sensors for air temperature, pressure, sun intensity and direction, amount of rain and quality of the air. Thus, each car is a kind of itinerant weather station. The collection and aggregation of this information from multiple vehicles could provide better weather forecasts than traditional services, due to the vehicular pervasivity over the global surface [3].

More in general, a proper exploitation of these sensors can be used in a opportunistic crowd-sensing fashion [4] to derive new real-time knowledge and to create new and exciting scenarios, services and business opportunities, in many different domains.

Mobility is another field where opportunistic crowd-sensing performed by vehicles is of particular relevance [5]. Indeed, having a deeper global insight on the current state of mobility situation, drivers could be supported by smarter Intelligent Transportation Systems (ITS). For example, drivers could be guided towards roads with less traffic and/or higher chances to find free parking spaces [6], based on real-time and forecasted traffic and parking data for the selected destination, also under exceptional circumstances and extreme events [7,8].

One of the key performance indicators of a crowd sensing system is the achievable spatio-temporal sensing coverage [4]. In the case of probe vehicles, this is mainly influenced by two factors: (I) the number of probes, and (II) the trajectories followed by probes. While there is a number of studies addressing the first issue (e.g.: [9–12]), to the best of our knowledge, there is only one paper investigating the impacts of trajectories on sensing quality [13]. Indeed, the huge literature on routing applied to smart mobility has mainly focused other problems, like eco-routing (e.g.: [14]) or dynamic car-pooling (e.g.: [15]), even if it is known that one of the current limitation with probe vehicles comes from the non-uniform distribution of vehicles over the road network. Indeed, drivers normally prefer a route leading to a shortest time path to a destination [13], which is an efficient delivery solution for logistics or passenger vehicles.

To address this issue, in this paper we propose a new probabilistic routing algorithm, corresponding to a variant of the standard $\mathbf{A}^*$ procedure, where, given the same source, destination, and road network, we compute potentially different routes, as long as the total travel distance is within a selectable threshold (in percentage over the total route). This is, in our opinion, a viable solution for improving the sensing coverage, without requiring a global coordination of the paths. The problem studied in this paper is relatively new and, to the best of our knowledge, no real robust approach has been proposed in the literature, since the solution proposed in [13] requires a global coordination of the vehicles.

We also performed a preliminary investigation of the algorithm, by exploiting a dataset of real taxi trajectories collected in San Francisco (USA) over five weeks. After cleansing the dataset, we selected more than 320,000 routes, and computed the potential improvements in terms of sensing coverage achievable by following the routes proposed by our algorithm.

The remainder of the paper is organized as follows. In Sect. 2, after some background information, the algorithm we propose to improve sensing coverage is presented. In Sect. 3 we describe the way we preliminary assessed the solution, while in Sect. 4 we present the preliminary computational results obtained by applying our strategy on a real scenario. Conclusions and final remarks are given in Sect. 5.

2 The Proposed Routing Algorithm

In this section we start by quickly recalling some basic concepts of the $\mathbf{A}^*$ algorithm, then we present the proposed solution.

2.1 The $\mathbf{A}^*$ Algorithm

An $\mathbf{A}^*$ search algorithm for the *single source-single destination* shortest path problem with nonnegative edge lengths (see among others [16]) focuses an "informed" search of a shortest path from a given source node s to a given destination node d in a weighted di-graph $G = (V, E)$ through the use of a heuristic function e. In more detail, it works with "modified" labels $l(v)$, $\forall v \in V$ given by $l(v) = d(v) + e(v, d)$, where $d(v)$ represents the classical label associated to v as in any classical shortest path problem method, while $e(v, d)$ expresses an estimate of the shortest distance from v to the destination node d.

An $\mathbf{A}^*$ algorithm is guaranteed to produce an optimal solution path (whenever a path from the source node to the destination node exists) if the heuristic function e is monotone, i.e., if $e(d, d) = 0$ and for all nodes x and y, it results that $e(x, d) \leq W(x, y) + e(y, d)$, where $W(x, y)$ is the shortest distance from x to y. It can be easily seen that if the heuristic function e evaluates to zero at each node, then $\mathbf{A}^*$ reduces to the Dijkstra's algorithm.

2.2   The Proposed Solution

The proposed work is based on the $\mathbf{A}_\varepsilon^*$ algorithm [17], presented by Judea Perl in 1982. Using not one, but two different heuristic functions to choose the expanding node, this routing algorithm could expand fewer nodes than the classical $\mathbf{A}^*$, at the cost of relaxing the optimality of the result.

The breakthrough of this $\mathbf{A}^*$ variant was that it gave a bound on the optimality relaxation, tied to a parameter ε, in which it guaranteed that the returned path would never be longer more than ε times the minimal one.

As the focus of the proposed work was identifying a route in a somewhat random manner, while limiting the maximum possible returned path length, $\mathbf{A}_\varepsilon^*$ proved to be the perfect base, even if not used as its authors originally intended.

The proposed algorithm, given the required source and destination of the route and a multiplicative coefficient ε, first finds in some way, for example using a standard implementation of $\mathbf{A}^*$, the length of the shortest path between the provided endpoints. If no such path exists, then the algorithm terminates, and returns an empty path itself. Otherwise, to better distribute the length of the different paths returned over time, a candidate lower bound $lBound$ is chosen randomly among n evenly spaced values comprised between the shortest path length l_P and $l_P \cdot \varepsilon$, the latter not included. The actual routing algorithm is now called and given as parameters the two endpoints, the lower and upper bounds ($lBound$ and $l_P \cdot \varepsilon$), and ε. If no path could be found that met the required bounds, a smaller lower bound is chosen, until either a path is found or $lBound = l_P$, which assures that at least the shortest path will be acceptable.

The inner path-finding algorithm, being based on $\mathbf{A}_\varepsilon^*$, works by creating, at every iteration, a list of nodes which are extremes to the best paths found until now, and, as such, are candidates for expansion: given an admissible and consistent heuristic function $h : V \to \mathbb{R}$ that estimates the distance of a node from the destination, a function $d : V \to \mathbb{R}$ that returns the length of the shortest path found until now between a node and the source, and their sum function $f(n) = h(n) + d(n)$, the list of candidates is comprised of all the nodes m such as

$$m \in \left\{ v \in V \mid f(v) \leq \min_{n \in V} f(n) \cdot \varepsilon \right\}. \tag{1}$$

After having done this, in $\mathbf{A}_\varepsilon^*$ a second heuristic $t : V \to \mathbb{R}$ is evaluated for every candidate node, and the one with the smallest image is chosen as the expanding node.

The basic version of the proposed algorithm differs from $\mathbf{A}_\varepsilon^*$ in this last operation, as it chooses the node to be expanded in a random manner. The different ways in which a node can be chosen randomly lead to different efficiency and effectiveness. In the preliminary experiments that were carried out, reported in this paper, every node in the list had a probability of being chosen equal to $\frac{2i}{N*(N+1)}$, where i is the index of the node in the total ascending ordering of the list by the value of f. Other possible strategies, such as giving the same probability of being chosen to every node, resulted in worse outcomes.

This first version, while completely working, was not the most efficient possible, and did not naturally offer the possibility of introducing a lower bound on returned paths. For these reasons, a bidirectional version was implemented and used for the experiments described later on in the paper. While being a pretty straightforward bidirectional version of an **A*** variant, it has been possible to simplify the termination criteria usually used in the latter because we were not interested in the shortest path, but in one of the many paths whose length was comprised between a lower and an upper bound: as soon as the algorithm encounters a fitting path, it terminates.

3   Experimental Design

In this section we describe the experimental protocol we adopted for the evaluation of the proposed algorithm, in terms of employed data, procedure and measures.

3.1   The Dataset

The taxi dataset was collected within the *Cabspotting* project [18], which aimed at the extraction of socio-economic properties of regions from the taxi patterns. Each taxi periodically provided information on its latitude and longitude, timestamp, and occupancy (1 = occupied, 0 = free) to a central server. The resulting dataset contains 11,219,955 GPS coordinates, collected from 536 vehicles of the *Yellow Cab* company, over 25 days in the San Francisco Bay Area, from 2008/05/17 until 2008/06/10.

The first challenging task was to obtain map matched trajectories, by aligning the sequence of GPS points contained in the FCD with the road network provided by *OpenStreetMap*. The median time gap between two consecutive GPS measures is 60 s, with time gaps ranging between 30 and 120 s, for 86% of all observations. Such a low frequency of the FCD collection, together with the intrinsic noise of the GPS, required using advanced map matching techniques. In more detail, since for each taxi we had FCD covering more than 3 weeks, the first step was to segment each taxi's data flow into a set of independent *trajectories*. A taxi trajectory T_r can be defined as a sequence of GPS points corresponding to a trip with the same passenger occupancy state. Each GPS point $p_i = (x_i, y_i, t_i)$ has a longitude x_i, latitude y_i, and a timestamp t_i. A trajectory T_r is thus a sequence of points $p_1 \rightarrow p_2 \rightarrow ... \rightarrow p_n$, where the state of passenger occupancy of the taxi is the same from p_1 to p_n. Thus, differently from other similar works (e.g. [19]), we are also considering taxi trajectories where the vehicle is not occupied, since clearly the sensing of parking spaces can be done also in these cases. We also split the sequence of points every time there was a time gap longer than 3 min between consecutive GPS points, on the assumption that the taxi was not operating in that time frame. Finally, from this set of trajectories, we discarded all those having less than 5 points and/or implausible speed between two consecutive points.

On these selected taxi trajectories, we applied the map matching algorithm described in [20], which is based on [21]. For each GPS point in a trajectory, candidate street segments are identified from the *OpenStreetMap* road network. Road segments between two consecutive GPS points projected on the map are identified by a shortest path search. In a following global optimization step, the sequence of candidate segments achieving the highest score, based on spatial and temporal criteria, is selected. As a result of this task, 3,371,552 GPS points were successfully matched to the *OpenStreetMap* road network.

Then, all trajectories containing a loop were removed, as well as all the trajectories containing 3 points or less. As a result, 420,790 potential taxi runs were identified.

3.2 Experimental Procedure

The experiments are thus executed on the previously described dataset. We checked each of these 420,790 traces, by re-running the standard $\mathbf{A}^*$ algorithm, given the respective sources and destinations of the considered paths. This allowed us on one hand to compute the results of the *baseline*, intended as the normal routes computed by the standard $\mathbf{A}^*$. On the other hand, this allowed us to further cleanse the dataset, by removing all the routes whose sequence of segments did not match the one provided by $\mathbf{A}^*$.

As a result, we kept 324,199 taxi runs, whose sources and destinations were used to evaluate the proposed algorithm. More in details, we choose six different configurations for ε, namely 1.05, 1.1, 1.15, 1.2, 1.25 and 1.3, thus permitting detours from 5% up to 30%. To minimize the bias due to the probabilistic nature of the algorithm, we run it 100 times for each route and each value of ε, and then we averaged across the 100 outcomes.

We computed the results for the following six attributes:

1. Total Length of the paths identified by the $\mathbf{A}^*$.
2. Total Length of the paths identified by the $\mathbf{A}^*_\varepsilon$.
3. Total Number of Segments covered by the $\mathbf{A}^*$.
4. Total Number of Segments covered by the $\mathbf{A}^*_\varepsilon$.
5. Road Segments covered only by the $\mathbf{A}^*$.
6. Road Segments covered only by the $\mathbf{A}^*_\varepsilon$.

4 Results and Discussion

In this section we present and discuss the obtained preliminary results we conducted. In particular, in Table 1 we report the summary statistics of the results obtained by the standard $\mathbf{A}^*$ algorithm and by the new $\mathbf{A}^*_\varepsilon$, on the couples (*Source*, *Destination*) on the taxi trips described in the previous section, with the six values of ε.

Let us note that the results of $\mathbf{A}^*_\varepsilon$ are intended as the average over the 100 runs. While the total distance covered by all the taxis using the $\mathbf{A}^*$ algorithm is 8,578 km (224,835 segments), this distance clearly varies with $\mathbf{A}^*_\varepsilon$ according to

Table 1. Summary statistics of the evaluation

Epsilon	1.05	1.1	1.15	1.2	1.25	1.3
A^* Length (km)	8,578					
A^*_ε Length (km)	9,262.45	9,714.80	10,127.56	10,549.33	10,942.40	11,282.39
A^* Segments	224,835					
A^*_ε Segments	253,739.6	268,653.3	279,566.8	289,261.4	297,708.1	304,507.2
Segments only A^*	5,599.53	6,866.58	8,181.95	9,338.1	10,451.37	11,462.82
Segments only A^*_ε	34,504.14	50,684.93	62,913.83	73,764.49	83,324.54	91,135.04

the value of ε. The same consideration holds also for the number of covered road segments. We can observe that the number of segments covered only by the A^* increases while changing ε. This is due to the fact that the bigger is ε, the more the A^*_ε looks for routes far from the standard version of the algorithm.

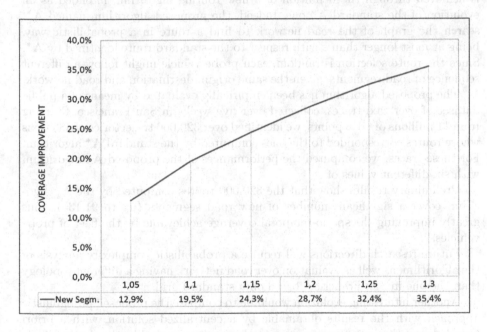

	1,05	1,1	1,15	1,2	1,25	1,3
—New Segm.	12,9%	19,5%	24,3%	28,7%	32,4%	35,4%

Fig. 2. Coverage increment, in %, due to the proposed algorithm, for different ε.

The most relevant information in the results is the number of new segments, covered only by A^*_ε. Indeed, this number represents the amount of new streets that might be potentially sensed by probe vehicles using our routing algorithm.

We also computed the relative improvement in terms of covered segments, defined as:

$$\frac{Segments\ Only\ A^*_\varepsilon - Segments\ Only\ A^*}{A^*\ Segments} \tag{2}$$

In Fig. 2 we report both in textual and graphical form these results.

From this figure we can see an interesting behavior, that, given an ε, the achievable relative improvement over the standard $\mathbf{A}^*$ is always greater than ε, in terms of new explored segments, and thus in sensing coverage of the probe vehicles.

5   Conclusion

A lot of research is currently being focused on exploiting the information that could be opportunistically crowd-sensed by the sensors of modern cars, to define new knowledge and/or services. Anyhow, an open issue with probe vehicles is the achievable spatio-temporal sensing coverage, since vehicles are not uniformly distributed over the road network, because drivers mostly select a shortest time path to destination.

In this paper we have presented the preliminary results of an on-going research towards maximizing the sensing coverage of probe vehicles. This goal is achieved through the definition of a new routing algorithm, intended as an evolution of the standard $\mathbf{A}^*$ one. Indeed, the proposed algorithm, named $\mathbf{A}^*_\varepsilon$, search the graph of the road network to find a route in a probabilistic way, being at most longer than ε with respect to the standard route identified by $\mathbf{A}^*$. Since the route selection is random, each probe vehicle might follow a different sequence of road segments, given the same origin, destination and road network.

The proposed algorithm has been empirically evaluated by means of a public dataset of over taxi traces, collected over five weeks in San Francisco. Starting from 11 millions of GPS points, we identified over 320,000 trajectories of 486 taxis whose routes corresponded to the ones computed by the standard $\mathbf{A}^*$ algorithm. For these traces, we computed the performances of the proposed $\mathbf{A}^*_\varepsilon$ algorithm, with six different values of ε.

Preliminary results show that the 320,000 routes computed by the $\mathbf{A}^*_\varepsilon$ algorithm cover a significant number of new road segments (up to 91,135), thus greatly improving the spatio-temporal coverage achievable by the fleet of probe vehicles.

Future research directions will require a probabilistic complexity analysis of the algorithm, as well as evaluation over road network having a different topology than the one in San Francisco, based on a standard grid.

Another interesting evolution would be to compare the proposed probabilistic approach with the results obtainable by a centralized solution, with a priori knowledge of the sensing coverage over the road network, like the one in [13].

References

1. Di Martino, S.: Discovering information from spatial big data for ITS. Intelligent Transport Systems (ITS): Past, Present and Future Directions, pp. 109–130 (2017)
2. Robert Bosch GmbH: Bosch Community-based parking (2016). http://www.bosch-mobility-solutions.com/en/connected-mobility/community-based-parking/. Accessed 27 June 2016

3. Petty, K., Mahoney III, W.: Enhancing road weather information through vehicle infrastructure integration. Transp. Res. Rec. J. Transp. Res. Board **2015**, 132–140 (2007)
4. Ganti, R.K., Ye, F., Lei, H.: Mobile crowdsensing: current state and future challenges. IEEE Commun. Mag. **49**(11), 32–39 (2011)
5. Wang, X., Zheng, X., Zhang, Q., Wang, T., Shen, D.: Crowdsourcing in ITS: the state of the work and the networking. IEEE Trans. Intell. Transp. Syst. **17**(6), 1596–1605 (2016)
6. Di Martino, S., Rossi, S.: An architecture for a mobility recommender system in smart cities. Procedia Comput. Sci. **98**, 425–430 (2016)
7. Kwoczek, S., Di Martino, S., Nejdl, W.: Stuck around the stadium? An approach to identify road segments affected by planned special events. In: 2015 IEEE 18th International Conference on Intelligent Transportation Systems (ITSC). IEEE, pp. 1255–1260 (2015)
8. Kwoczek, S., Di Martino, S., Nejdl, W.: Predicting and visualizing traffic congestion in the presence of planned special events. J. Vis. Lang. Comput. **25**(6), 973–980 (2014)
9. Bock, F., Martino, S.D., Sester, M.: What are the potentialities of crowdsourcing for dynamic maps of on-street parking spaces? In: Proceedings of the 9th ACM SIGSPATIAL International Workshop on Computational Transportation Science. ACM, pp. 19–24 (2016)
10. Bock, F., Attanasio, Y., Di Martino, S.: Spatio-temporal road coverage of probe vehicles: a case study on crowd-sensing of parking availability with taxis. In: Bregt, A., Sarjakoski, T., van Lammeren, R., Rip, F. (eds.) GIScience 2017. LNGC, pp. 165–184. Springer, Cham (2017). https://doi.org/10.1007/978-3-319-56759-4_10
11. Mathur, S., Jin, T., Kasturirangan, N., Chandrasekaran, J., Xue, W., Gruteser, M., Trappe, W.: ParkNet: drive-by sensing of road-side parking statistics. In: Proceedings of 8th International Conference on Mobile Systems, Applications, and Services, pp. 123–136. ACM, New York (2010)
12. Bock, F., Di Martino, S.: How many probe vehicles do we need to collect on-street parking information? In: 2017 5th IEEE International Conference on Models and Technologies for Intelligent Transportation Systems (MT-ITS), pp. 538–543. IEEE (2017)
13. Masutani, O.: A sensing coverage analysis of a route control method for vehicular crowd sensing. In: 2015 IEEE International Conference on Pervasive Computing and Communication Workshops (PerCom Workshops), pp. 396–401. IEEE (2015)
14. Boriboonsomsin, K., Barth, M.J., Zhu, W., Vu, A.: Eco-routing navigation system based on multisource historical and real-time traffic information. IEEE Trans. Intell. Transp. Syst. **13**(4), 1694–1704 (2012)
15. Di Martino, S., Giorio, C., Galiero, R.: A rich cloud application to improve sustainable mobility. In: Tanaka, K., Fröhlich, P., Kim, K.-S. (eds.) W2GIS 2011. LNCS, vol. 6574, pp. 109–123. Springer, Heidelberg (2011). https://doi.org/10.1007/978-3-642-19173-2_10
16. Lugar, G.F., Stubblefield, W.A.: Artificial Intelligence Structures and Strategies for Complex Problem Solving. Addison Wesley Longman, Harlow (1998)
17. Pearl, J., Kim, J.H.: Studies in semi-admissible heuristics. IEEE Trans. Pattern Anal. Mach. Intell. **4**, 392–399 (1982)
18. Piorkowski, M., Sarafijanovic-Djukic, N., Grossglauser, M.: CRAWDAD dataset EPFL/mobility, 24 February 2009. http://crawdad.org/epfl/mobility/20090224. Accessed Feb 2009

19. Chen, C., Zhang, D., Ma, X., Guo, B., Wang, L., Wang, Y., Sha, E.: CrowdDeliver: planning city-wide package delivery paths leveraging the crowd of taxis. IEEE Trans. Intell. Transp. Syst. **18**(6), 1478–1496 (2017)
20. Axer, S., Pascucci, F., Friedrich, B.: Estimation of traffic signal timing data and total delay for urban intersections based on low-frequency floating car data. In: Proceedings of the 6th Mobility, TUM 2015 (2015)
21. Lou, Y., Zhang, C., Zheng, Y., Xie, X., Wang, W., Huang, Y.: Map-matching for low-sampling-rate GPS trajectories. In: ACM SIGSPATIAL GIS 2009. ACM, November 2009. https://www.microsoft.com/en-us/research/publication/map-matching-for-low-sampling-rate-gps-trajectories/

Generation of Web-Based GIS Applications Through the Reuse of Software Artefacts

Alejandro Cortiñas(✉)(iD), Miguel R. Luaces(iD), Oscar Pedreira(iD), and Ángeles S. Places(iD)

Laboratorio de Bases de Datos, Universidade da Coruña, A Coruña, Spain
{alejandro.cortinas,luaces,oscar.pedreira,asplaces}@udc.es

Abstract. This demo shows the automatic generation of different web-based geographic information systems using a tool based on software product lines engineering. These systems are variant regarding the data domain they can manage and the functionalities they provide. Although the products are different, the set of assets that implement the GIS-related functionalities is the same. These assets are assembled together by our tool depending on the particular requirements of each products. In the demo, we show the behaviour of both the tool generating the products and the products themselves.

Keywords: Web-based geographic information systems
Software product lines engineering · Software engineering

1 Introduction

Developing web-based geographic information systems (GIS) is a costly task. In addition to the traditional intricacy of designing and implementing any information system, the geographic component adds an extra level of difficulty: the system has to handle a different kind of data with its own functionality that is specially complex if we compare it to alphanumeric data processing.

The current standardization of the domain makes the design of GIS-related software artefacts easier than ever. In fact, it is possible to define all the data and procedures required by a particular web-based GIS product using a domain-specific model, develop reusable software artefacts that implement this model, and define and define and build web-based GIS products by specifying an instance of the model and combining the software artefacts accordingly. There are some previous works in this direction: [4] describes a tool which generates a

This work has been funded by Xunta de Galicia/FEDER-UE CSI:ED431G/01; GRC: ED431C 2017/58. MINECO-CDTI/FEDER-UE CIEN LPS-BIGGER: IDI-20141259; INNTERCONECTA uForest: ITC-20161074. MINECO-AEI/FEDER-UE Datos 4.0: TIN2016-78011-C4-1-R; Flatcity: TIN2016-77158-C4-3-R; ETOME-RDFD3: TIN2015-69951-R. EU H2020 MSCA RISE BIRDS: 690941.

© Springer International Publishing AG, part of Springer Nature 2018
M. R. Luaces and F. Karimipour (Eds.): W2GIS 2018, LNCS 10819, pp. 11–14, 2018.
https://doi.org/10.1007/978-3-319-90053-7_2

web-based GIS with a fixed set of features adapted to a particular data domain; instead, [1] shows a set of web-based GIS prodcuts with some variable features, but sharing all of them the same data domain. Both approaches share one point: they focus on web applications, which gives us an idea of how important are web-based GIS nowadays.

We created a tool that can generate web-based GIS products where both the data domain and the functionalities are variant. This tool is based on a well-known methodology on software engineering: software product line engineering (SPLE). This methodology is based on reuse strategies and mass-customization, aiming to achieve the automatic generation of software in a industrial way. SPLE has been applied to many domains to reduce the cost and time to market of new software applications and, at the same time, to improve the quality of their code. In [3], we defined a SPLE for the generation of web-based GIS using a methodological process that includes the analysis of existing GIS applications and architectures, the standards defined in this domain and the knowledge of a software company with a certain grade of experience.

In this demo, we show the tool that we designed and implemented a tool that satisfies the requirements defined in [3] and that it is able to generate web-based GIS products having the most common functionalities for these systems with different levels of complexity. The users of our tool are also able to specify the data model that the new application has to handle. Therefore, we can create web-based GIS applications for any data domain. The technical description of this tool is described in [2].

2 Reusing Software to Generate GIS Applications

In [3], we identified and described an exhaustive set of requirements and features for web-based GIS products. A table summarizing the requirements can be found as supplementary material[1]. Similarly, the full table of features that the Software Product Line platform must provide in order to is available as supplementary material[2]. Some of the features are standard for any web application, such as every feature related to *User Management*. The other features focus on providing functionalities common to web-based GIS-products existing in the market.

Our tool can create products that implement GIS functionalities over a particular data model schema. For example, the component to *import shapefiles* can be used both by an application that handles warehouse logistics or by one that manages geolocated sensors. Figure 1 shows a screenshot of the component being use to import some *road* data into an application.

Another set of feature that our tool provides is *importation, visualization and management of raster* layers (see Fig. 2). Figure 2a shows the *layer management* feature that can be used to change the layers present in the map viewer, to change their style or opacity, to centre the map in the bounding box of a layer, or to download the data of a layer. In Fig. 2b, the *map viewer* is shown with

---

[1] Web-based GIS requirements: http://lbd.udc.es/webgis-spl/requirements.pdf.
[2] Web-based GIS feature list: http://lbd.udc.es/webgis-spl/featurelist.pdf.

Fig. 1. Shapefile importation feature

(a) Layer manager feature

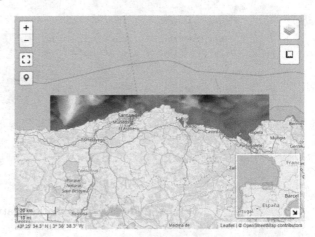

(b) Map with a imported raster

Fig. 2. Raster importation and viewing

a raster layer of the Cantabrian Sea loaded, with some tools such as *context information, user geolocation, zoom-to-window button* or *measure toolbox*.

Our tool is also able to generate products with some advanced features implemented, such as network tracing or route calculation. Specifically, we provide a REST service that can be called from our map viewer, using a simple interface, or from any other client application connected to the GIS. For example, we could calculate the places accessed from the centre of Madrid in a 30 min travel by car Fig. 3a or the route to go from Lugo to A Coruña Fig. 3b.

In Fig. 4 we show the *address geolocation* component that be used to compute the geographic coordinates of an address associated to an entity of the data model. All the features described in this paper are some examples of the great grade of customization achievable with our tool.

(a) Network tracing feature   (b) Route calculation feature

Fig. 3. Connectivity related features

Fig. 4. Address geolocalization feature

References

1. Buccella, A., Cechich, A., Pol'la, M., Arias, M., del Socorro Doldan, M., Morsan, E.: Marine ecology service reuse through taxonomy-oriented SPL development. Comput. Geosci. **73**, 108–121 (2014)
2. Cortiñas, A., Luaces, M., Pedreira, O., Places, Á.: Scaffolding and in-browser generation of web-based GIS applications in a SPL tool. In: Proceedings of the 21st International Systems and Software Product Line Conference (2017)
3. Cortiñas, A., Luaces, M., Pedreira, O., Places, Á., Pérez, J.: Web-based geographic information systems SPLE: domain analysis and experience report. In: Proceedings of the 21st International Systems and Software Product Line Conference. ACM, Sevilla (2017)
4. Devoine, P.A., Moisuc, B., Gensel, J.: GENGHIS: an environment for the generation of spatiotemporal visualization interfaces. In: Innovative Software Development in GIS. Wiley (2013)

Storing and Clustering Large Spatial Datasets Using Big Data Technologies

Alejandro Cortiñas(✉), Miguel R. Luaces, and Tirso V. Rodeiro

Laboratorio de Bases de Datos, Universidade da Coruña, A Coruña, Spain
{alejandro.cortinas,luaces,tirso.varela.rodeiro}@udc.es

Abstract. In this paper we present the architecture of a system to store, query and visualize on the web large datasets of geographic information. The architecture includes a component to simulate a large number of drivers that report their position on a regular basis, an ingestion component that is generic and can acommodate three different storage technologies, a query component that aggregates the results in order to reduce the query time and the data transfered, and a web-based map viewer. In addition, we define an evaluation methodology to be used to benchmark and compare different alternatives for some components of the system, and we validate the architecture with experiments using a dataset of 40 million locations of drivers.

Keywords: Spatial big data · Web-based GIS · Software architectures

1 Motivation

Current technology makes the real-time collection of massive volumes of geographic information feasible. The computing power of current mobile devices is similar to the one of a desktop computer from the last decade. They can be used to measure different variables such as the geographic position using a GPS receiver, or the user activity using an accelerometer. *Mobile Workforce Management (MWM)* technologies will benefit especially from the information collected using mobile devices, and they are gaining attention because they are used by companies to manage and optimize the task scheduling of their workers and to improve the performance of their business processes [3]. Hence, Mobile Workforce Management is useful to detect patterns in the past activity of workers, or to predict trends that can improve the future scheduling. For example, Mobile Workforce Management can be used by a company to detect which tasks are costly for the company, or to ensure that it has the lowest number of active employees at any time of the day.

This work has been funded by Xunta de Galicia/FEDER-UE CSI: ED431G/01; GRC: ED431C 2017/58. MINECO-CDTI/FEDER-UE CIEN LPS-BIGGER: IDI-20141259; INNTERCONECTA uForest: ITC-20161074. MINECO-AEI/FEDER-UE Datos 4.0: TIN2016-78011-C4-1-R; Flatcity: TIN2016-77158-C4-3-R. EU H2020 MSCA RISE BIRDS: 690941.

M. R. Luaces and F. Karimipour (Eds.): W2GIS 2018, LNCS 10819, pp. 15–24, 2018.
https://doi.org/10.1007/978-3-319-90053-7_3

Datasets produced by mobile sensing and Mobile Workforce Management technologies are large and complex. As an example, consider a carsharing service like car2go. If each car produces a GPS position every 10 s (64 bytes considering a device id, a timestamp, three geographic coordinates, speed, bearing, and accuracy), it generates 552,900 bytes of data every day. Considering that car2go in Madrid has 500 vehicles [1], it would require over than 263 MB of storage each day. Larger systems, such as the taxi fleet (over 16,000 licenses in the region of Madrid), or the inclusion of additional sensor data (such as accelerometer data) would produce larger datasets.

Although the datasets are large, storage space is affordable and it does not seem impossible to store all this information, but querying and visualizing these datasets is a still difficult task. Current web-based GIS technology cannot deal with this problem. Data management technologies have been working during the last years to support horizontal scaling and distributed processing [2,5,7,8]. Therefore, storing and querying large geographic datasets can be achieved. However, middleware software such as map servers and visualization software such as javascript-based map viewers are not adapted to large datasets and distributed processing. Consider, for example the map server Geoserver [6]. It supports querying multiple data sources, but the support for MongoDB is quite recent and there is no support for other NoSQL technologies. Similarly, consider the map viewer Leaflet [4]. It supports different types of data layers but it includes plugins to aggregate large datasets to avoid cluttering the map display. However, the aggregation is performed on the client side, thus requiring large datasets to be transferred over the network and to be processed in the web browser. Hence, in order to support the visualization of large geographic datasets, middleware components and map viewers must support querying and aggregating geographic data using distributed processing systems.

In this paper, we present the architecture of a system to store, query and visualize on the web large datasets of geographic information (see Sect. 2). Given that selecting the most appropriate technology to support these usage scenarios is complicated, we have started to define an evaluation methodology to be used to benchmark and compare different alternatives for some components of the system (see Sect. 3). Therefore, the architecture includes a system to simulate the real load on one of these systems by simulating a large number of drivers that circulate through a road network and report their position to the server on a regular basis. In addition, the system also provides a solution to feed different storage subsystems in a simple way with the same set of data so that they can be tested under the same load conditions and evaluate their performance with the same queries. Finally, we also show preliminary results in which we use the proposed system to store, query and visualize 40 million locations of drivers in a relational spatial database management system (PostgreSQL+PostGIS), a NoSQL database management system (MongoDB) and a Big Data tool for storing and querying OLAP data (Druid) (see Sect. 4). Finally, we discuss the results and propose future work (see Sect. 5).

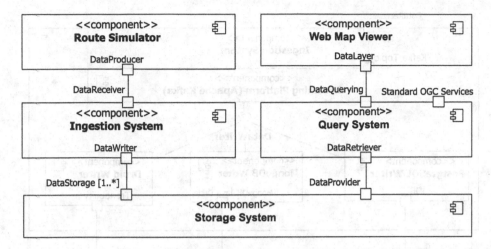

Fig. 1. System architecture

2   Proposed Architecture

Figure 1 shows the main components of the system architecture. The `Route Simulator` component runs on the client side of the system and it simulates a collection of simultaneous drivers. Each driver starts at a random position in the road network, it calculates a route to a random destination, and it generates positions along the sections of the route with the specified periodicity assuming a random speed expressed as a percentage of the maximum allowed speed. For example, a driver can generate positions every second assuming that it always circulates at 80% of the maximum speed of the road, while another driver can generate positions every five seconds running at 105% of the maximum speed of the road. The component `Ingestion System` runs on the server side and it is responsible for collecting the data in real time. Its implementation is generic so that it is very easy to implement extensions to the component that behave differently simply by implementing a Java interface. The component `Storage System` is in charge of storing the data in a database management system.

Regarding the querying and visualization functionality of the architecture, final users interact with the `Web Map Viewer` component, which is based on a standard web-based map viewer. Additionally, considering that the datasets that have to be visualized are extremely large, the information in these datasets is clustered depending on the zoom level in order to avoid cluttering the map display. The process of query solving and data clustering is performed on the server side by the `Query System` component, which receives queries sent by the `Web Map Viewer` component, delegates them on the `Storage System` component, and sends the results back to the client.

Figure 2 shows a detailed view of the ingestion architecture components. The components with a grey background are used without modifications. The entry point of the `Ingestion System` component is a messaging system (Apache

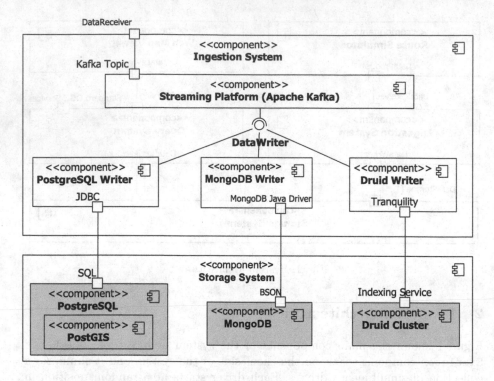

Fig. 2. Detailed ingestion architecture

Kafka) that is used to decouple the task of receiving large collections of data from the storage. The `Streaming Platform` component consumes the Kafka events and forwards them to different storage systems. Figure 3 shows an example of an event received by the `Ingestion System` component. It consists of information regarding the worker id, the GPS position of the worker, the timestamp of the position, and additional information in JSON format that is specific of the particular domain for which the architecture is being used.

The communication with the `Storage System` component is managed by a component that implements the generic `Data Writer` interface. We have currently implemented three alternatives: one that stores the events in Post-gres+PostGIS (the component `PostgreSQL Writer`), another one that stores the data in MongoDB (the component `MongoDB Writer`), and another that stores the data in Druid (the component `Druid Writer`).

Given that the purpose of the data stored in the system is the interactive visualization by the user in a map viewer, and that having a fluid visualization in the client side is a very important requirement, it is necessary to aggregate the points on the server side of the application and send to the client side only the result of the aggregation. That is, instead of transferring large collections of individual geographic points, we want to transfer a smaller collection of clustered points and the number of events that are added under that point. Furthermore, considering

```
1  {
2    "worker_id": 3,
3    "position": {
4      "x": -8.0041161,
5      "y": 43.18902,
6      "z": 517,
7      "speed": 32.48,
8      "bearing": 83.6,
9      "accuracy": 4.5509996
10   },
11   "timestamp": 1513763866,
12   "data": {...additional data in json format...}
13 }
```

Fig. 3. Example of an event received by the `Ingestion System` component

that the user will be able to define specific spatial and temporal ranges for the set of events that have to be retrieved to visualize by means of zoom and pan operations in a map and a time range control, precomputed clusters cannot be used because the variation among queries is too large. The simplest alternative is to perform the query *get all points in the range (xmin, ymin, tmin) - (xmax, ymax, tmax)* and apply an aggregation algorithm on the result, but it is a costly solution in terms of computation requirements. Instead, we propose to preprocess each event when it is inserted into the system and add versions of the geographic point with decreasing precision. For example, if each point is represented in a geographic coordinate system in which the coordinates represent longitude and latitude in degrees, a value with an accuracy of 9 decimals represents a maximum of 1 mm on the surface of the Earth. Storing 7 additional versions of the same geographical point with 7 different precisions (between 2 and 8 decimals) makes the process of aggregation as simple as grouping the events by equal values of coordinates and counting the number of elements. Therefore, the components implementing the `Data Writer` interface are also in charge of computing and storing the additional versions of the geographic point for each event.

Figure 4 shows a detailed view of the querying architecture components. The components with a gray background are used without modifications. The same design pattern that was used in the `Ingestion System` component is used to manage the communication with the `Storage System`. A generic interface was defined (`DataRetriever`) and components implementing this interface are responsible for parsing the query and communicating with the `Storage System` component. We implemented three alternatives: Postgres+PostGIS (`PostgreSQL Retriever`), MongoDB (`MongoDB Retriever`), and Druid (`Druid Retriever`).

The data retrieving components are used in two alternative use cases. In the first use case, we support direct visualization of the data in the `Web Map Viewer` component using `Leaflet` and two components that manage the communication: a client-side component called `LeafletDataLayer` that implements the Layer interface of Leaflet, and a server-side component called `Leaflet Backend` that receives queries and delegates them to the appropriate data retrieving component. In the second use case, we implemented an extension to `Geoserver`, a popular map server component. This extension (named `Geoserver Backend`) acts as an adapter between Geoserver and the data retrieving components. Whereas the first scenario avoids middleware layers and performs faster, the second scenario provides applications with standard OGC services such as WMS, WFS and Filter Encoding. Figure 5 shows an example of a user interface using the Web Map Viewer component. The user interface displays a map with clusters that represent the amount of events that are stored in the database in a time range specified by the users.

3   Evaluation Methodology

In order to validate the architecture that we have proposed in Sect. 2, we have identified three key aspects that have to be evaluated and that lead to three research questions:

- *Research question 1 (RQ1). Which of the candidate storage technologies provides a faster answer to aggregation queries?* The selection of a storage technologies cannot be based only on how fast is able to answer a specific type of query because there may be many other requirements (e.g., transaction support, horizontal scaling, etc.). However, being able to answer the aggregation queries that are the basis of the architecture is a very important requirement with a high weight on the decision.
- *Research question 2 (RQ2). Can we achieve a constant time in aggregation queries using different versions of each geographic point?* In Sect. 2, we proposed a simple approach that can be used to improve aggregation queries using multiple versions of each geographic point. It is clear that this approach requires additional storage space, but it is not clear whether it will be possible to achieve a constant time answering queries.
- *Research question 3 (RQ3). Which query predicate performs faster, the distance based or the geometry based?* All storage systems provide two alternatives to solve range queries: one based on a distance predicate, which uses mathematical operations, and another one based on a bounding box predicate, which uses geometric operations. It is important to discover which alternative of the queries performs faster in order to use it in queries.

To generate test datasets, we have run the `Route Simulator` component in a desktop computer (Intel Core i7-3770, 4 cores, 3.40 GHz, 8 GB of RAM). The use of Kafka asynchronously as the entry point of the `Ingestion System` component makes the client very light and it can generate events for 1000 simultaneous drivers with hardly any load for the system. The server side was run on a machine (Intel Core i5-4440, 4 cores, 3.10 GHz, 14 GB of RAM) that hosted all the server-side components of the architecture (`Ingestion System`, `Storage System`, and `Query System`). Only one storage technology was running simultaneously (either Postgres+PostGIS, MongoDB or Druid) in order to avoid resource allocation

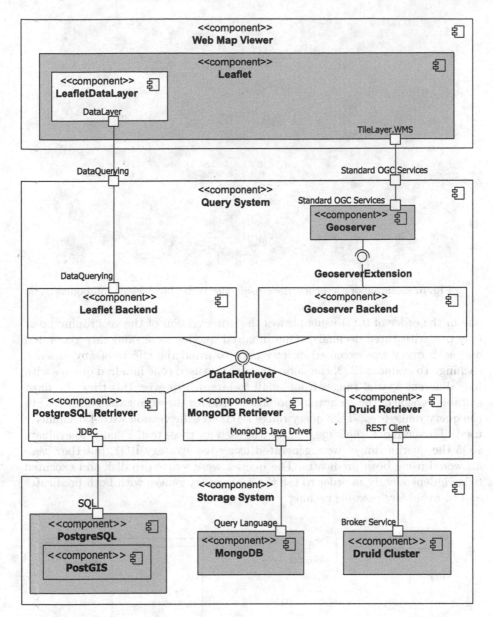

Fig. 4. Detailed querying architecture

competitions. The system has behaved in a stable manner and has allowed to generate two datasets: one with 3.2 million points in 3 h (*small dataset*), and another with 40.7 million points in 24 h of continuous execution (*large dataset*).

In order to evaluate RQ1, for each candidate storage, we generated queries with four different spatial ranges (from small ones in the order of 0.0001° to large

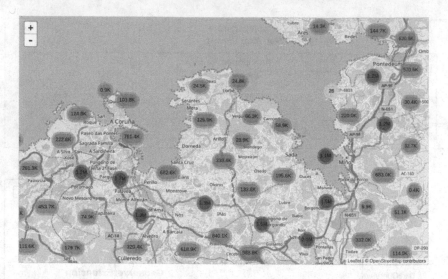

Fig. 5. Example of a user interface using the Web Map Viewer component

one in the order of 0.1°) using always the same version of the geographic point (the one with three decimals). One hundred queries were randomly generated and each query was executed exactly once to avoid the effects of any possible caching. To evaluate RQ2, the same strategy was used (one hundred queries with four different spatial ranges from small to large). However, this time the most suitable version of the geographic point was used for the aggregation according to the query range (e.g., if the query range is 0.0001° the version with 4 decimals is used). Furthermore, the large dataset was used for these test. Finally, to evaluate RQ3 the queries ranges were generated using the strategy RQ1, but they were answered using both predicates. The queries were written to disk and executed twice independently in order to use the same query ranges with both predicates and to avoid any possible caching.

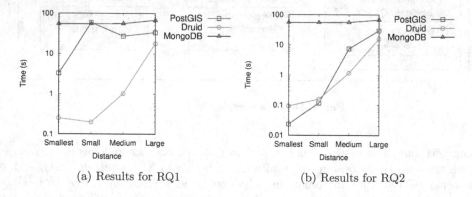

(a) Results for RQ1   (b) Results for RQ2

Fig. 6. Experiment results for RQ1 and RQ2

4   Experiments and Results

Figure 6a shows the results of the experiment performed to evaluate RQ1. The horizontal axis represents the different spatial ranges from small to large, and the vertical axis represents the average time in seconds to answer 100 queries using a logarithmic scale. It can be seen that Druid outperforms Postgres+PostGIS, which in turn outperforms MongoDB. The results indicate that Druid is the best option for the storage technology.

Figure 6b shows the results of the experiment performed to evaluate RQ2 using the same axis. The results are similar, even though the performance of Postgres+PostGIS improves yielding results comparable to Druid. However, the time required to answer queries increases with larger spatial ranges instead of remaining constant. The result indicate that our approach cannot be used to achieve a constant query time.

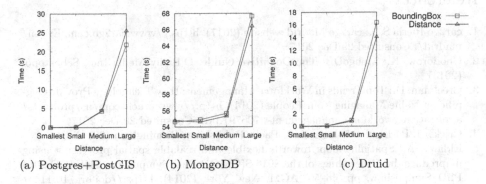

(a) Postgres+PostGIS   (b) MongoDB   (c) Druid

Fig. 7. Experiment results for RQ3

Figure 7 shows the results of the experiment performed to evaluate RQ3. The horizontal axis represents the different spatial ranges from small to large, and the vertical axis represents the average time in seconds to answer 100 queries using a different linear scale in each figure. It can be seen that the performance of both predicates (distance and bounding box) is similar. However, in Postgres+PostGIS the bounding box predicate outperforms the distance predicate, whereas in MongoDB and Druid the results are the opposite. The results indicate that the predicate used in the queries does not have a significative effect on the performance.

5   Conclusions and Future Work

We have presented in this paper the architecture of a system to store, query and visualize on the web large datasets of geographic information. We also have started to define an evaluation methodology to benchmark and compare different alternatives for some components of the system. We also show preliminary results in which we use the proposed system to store, query and visualize 40 million locations of drivers in PostgreSQL+PostGIS, MongoDB, and Druid. We validated the system and we can conclude that Druid is the most promising technology. We also evaluated our approach to achieve a constant time in aggregation queries and we can concluded that it is not successful. We believe that the problem with the approach is that truncating a decimal means that one in a level of aggregation represents one hundred points in the next level of aggregation. Thus, the difference between the different levels of aggregation is too high. Finally, we evaluated whether the specific predicate used in the queries is relevant and we found out that there is no significative difference between them.

As future work, we are currently working on evaluating which technology scales better in a cluster, and new approaches to achieve constant query time in aggregation queries.

References

1. car2go Iberia S.L.: car2go Madrid website (2017). https://www.car2go.com/ES/en/madrid/. Consulted 29 Dec 2017
2. Chodorow, K.: MongoDB: The Definitive Guide. O'Reilly Media Inc., Sebastopol (2013)
3. Creelman, D.: Top Trends in Workforce Management: How Technology Provides Significant Value Managing Your People (2014). http://www.oracle.com/us/products/applications/workforce-management-2706797.pdf. Consulted 29 Dec 2017
4. Crickard, P.: Leaflet.Js Essentials. Packt Publishing, Birmingham (2014)
5. Eldawy, A.: Spatialhadoop: towards flexible and scalable spatial processing using mapreduce. In: Proceedings of the 2014 SIGMOD PhD Symposium, SIGMOD 2014 PhD Symposium, pp. 46–50. ACM, New York (2014). https://doi.org/10.1145/2602622.2602625
6. Henderson, C.: Mastering GeoServer. Packt Publ., Birmingham (2014)
7. Obe, R.O., Hsu, L.S.: PostgreSQL: Up and Running a Practical Introduction to the Advanced Open Source Database, 2nd edn. O'Reilly Media Inc., Sebastopol (2014)
8. Yang, F., Tschetter, E., Léauté, X., Ray, N., Merlino, G., Ganguli, D.: Druid: a real-time analytical data store. In: Proceedings of the 2014 ACM SIGMOD International Conference on Management of Data, SIGMOD 2014, pp. 157–168. ACM, New York (2014). https://doi.org/10.1145/2588555.2595631

An In-depth Analysis of CUSUM Algorithm for the Detection of Mean and Variability Deviation in Time Series

Rayane El Sibai[1]([⊠]), Yousra Chabchoub[1], Raja Chiky[1], Jacques Demerjian[2], and Kablan Barbar[2]

[1] LISITE Laboratory, ISEP, 92130 Issy Les Moulineaux, France
rayane.el_sibai@etu.upmc.fr, {yousra.chabchoub,raja.chiky}@isep.fr
[2] LARIFA-EDST Laboratory, Lebanese University, Fanar, Lebanon
{jacques.demerjian,kbarbar}@ul.edu.lb

Abstract. Assessing and enhancing data quality in sensors networks is an important challenge. In fact, data recorded and sent by the sensors are often dirty, they contain noisy, erroneous and missing values. This can be due to many reasons such as sensor malfunction, uncalibrated sensor and low battery of the sensor or caused by external factors such as climatic conditions or interference. We focus in this paper on change detection in the time series data issued from sensors. We particularly address slow and gradual changes as they illustrate sensor calibration drift. For this purpose, we provide in this paper an in-depth analysis and improvement of the well known Cumulative Sum (CUSUM) control chart algorithm, as it is well adapted to small shifts detection. First, we discuss the choice of the different parameters of CUSUM in order to optimize its results based on the trade-off between the false positives and the Average Run Length (ARL_δ) needed by the algorithm to detect a process mean shift of δ. A study of the variability of the Run Length (RL_δ) is provided using simulation. Then, we introduce two improvements to the CUSUM algorithm: we propose an efficient method to estimate the starting point and the end point of the mean shift. Moreover, we adapt CUSUM to detect not only process mean deviation but also process variability deviation. All these improvements are validated by simulation and against real data stream issued from water flow-meters.

Keywords: CUSUM algorithm · Change point detection
Process variability · Sensors networks · Data quality

1 Introduction

Data streams have significantly increased with the development of the Internet of Things (IoT) and the emergence of connected objects. Large and heterogeneous collections of data are continuously generated by these objects at a high rate. They are issued from the activity of different organizations belonging to

© Springer International Publishing AG, part of Springer Nature 2018
M. R. Luaces and F. Karimipour (Eds.): W2GIS 2018, LNCS 10819, pp. 25–40, 2018.
https://doi.org/10.1007/978-3-319-90053-7_4

various domains such as healthcare, financial services, social networks, logistics and transport, and public administration.

Detecting online a change in the data stream is an important issue as it can have several interpretations depending on the application domain. In [1], the authors showed that an abrupt increase in the number of destination ports in IP traffic is an efficient criterion to detect port scan attacks. In [2], an analysis of data streams issued from a multi-temporal satellite is provided. It aims to detect land use and land cover changes. Such changes can be explained by human activities, natural conditions and development activities.

Several change detection methods have been designed by the statistics research community. According to [3], these methods can be classified into two categories. First, we distinguish algorithms designed to detect changes in the scalar parameters of an independent sequence. In this category, one can cite Shewhart control charts, geometric moving average control chart, CUSUM algorithm and Bayes-type algorithms. The second category concerns algorithms that are adapted to more complex changes such as non-additive changes in multidimensional signals. Among these methods, one can mention AR/ARMA models and the likelihood ratio. Several criteria can be considered to compare all these methods. The comparison can be based on the tolerance to false positives (false alarms), the response time of the algorithm (Run Length to detect the change), or the adaptation to progressive (or small) changes. The choice of the suitable change detection method depends on the application field and the specificity of the targeted error or change.

Statistical Process Control (SPC) is a set of techniques and tools for monitoring the quality of a process. One of the main used tools are the control chart algorithms. The statistical control chart concept was first introduced by *Walter A. Shewhart* from the Bell Telephone Laboratories in 1924. Control chart algorithms are particularly used to monitor the process stability over time by detecting a change in its parameters. The most known control chart algorithms are Shewhart, exponential/geometric moving average and CUSUM control charts. The type of the control chart depends on the number of process characteristics to be monitored. The univariate control chart is a chart of one quality characteristic, while the multivariate control chart represents more than one quality characteristic.

Control charts distinguish the habitual from the non-natural variation of the process. They signal an alarm when the process presents a suspicious deviation from a standard behavior. This deviation is defined based on two given thresholds called control limits. In general, the chart contains three elements: the plotted data corresponding to the process itself, the control limits, and the process average. By comparing the plotted data to these control limits, a decision about the stability of the process is inferred. The process is said in-control when it is stable and out-of-control otherwise. As long as the process remains in-control, the data points fall within the control limits. If a data point falls outside the control limits, we consider that the process is out-of-control. An investigation is thus needed to find and eliminate the cause(s) of the occurred change.

One of the most commonly used control chart algorithms to control the process average is the Shewhart. It evaluates the state of the process based only on the information concerning the last observation of the process while ignoring the past observations. It is efficient for the detection of large shifts, as it has a shorter response time than CUSUM in this case. However, it is insensitive to small changes. CUSUM chart overcomes this problem by using the current observed data and the historical observed data. It cumulates the impact of small deviations over time what enables it to detect small shifts in the process mean. Another control chart designed to detect a shift in the process average is the Exponentially Weighted Moving Average (EWMA) [4]. It is based on a weighted average that is updated for each received item of the stream. This moving average takes into account the current item and all the observed data. More importance (higher weight) is assigned to recent data.

The main performance measure of a control chart is the Average Run Length (ARL). Run Length (RL) is the number of observations required by the control chart algorithm to signal an alarm. When the process is in-control, the RL refers to false positives rate, and in the case of change, it qualifies the response time (reactivity) of the algorithm. RL will be discussed in details in the next section.

In this paper, we focus on the cumulative sum CUSUM algorithm, a well known control chart algorithm, proposed by Page in 1954 [5] and used to detect a shift in the process parameters. It aims to monitor the variation of the average of a process, and has the ability to detect small shifts (less than 1.5σ) from the expected average (see [6] for more details). The cumulative sum CUSUM control chart was initially proposed by Page in 1954 [5], and has been addressed in several research studies, such as [7–11].

Since the response time of CUSUM to detect a change depends on the parameters chosen beforehand, we study the impact of these input parameters on the speed of the detection. After that, we propose two improvements in order to enhance the estimation of the start time and end time of each detected error. Finally, we adapt CUSUM to detect not only process mean deviation but also process variability deviation. We validate our approach by simulation and on real data concerning water consumption with some stuck-at errors.

The explored data is issued from a flow-meter delivering water to a particular sector (geographical area) in a big French city. The flow-meters measure the flow delivered to this sector. The flow measurements for a given flow-meter are very variable. Indeed, they depend on the other flow-meters supplying the associated sector. For example, a given source associated with a particular flow-meter may suddenly stop supplying water to a sector. The latter will be delivered by its other associated peripheral sources.

Actually, the big French city is divided into several sectors, where many sensors have been deployed by the water operators on the periphery of each sector. The sensors periodically measure and send to the central monitoring system many water-related observations such as the flow, pressure, and chlore. Each of these deployed sensors has two categories of attributes: spatial attributes and non-spatial attributes. The spatial attributes depict the geographical location of

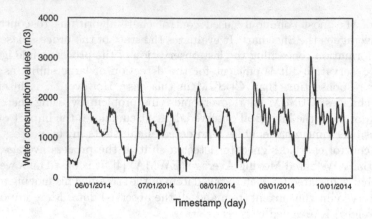

Fig. 1. Volume of the water consumed by the sector, over time

the sensor: latitude and longitude, while the non-spatial attributes include the name, ID, and the record observations of the sensor.

The data recorded by the sensors are structured data streams and have both spatial and temporal characteristics. In addition to the geographical position of the sensor, each record observation is composed of two fields: the timestamp designating the recording date of the measure, and the value of the measure. These data are regularly generated by the sensors with a frequency of one observation each 15 min.

Figure 1 shows the volume of the consumed water of a specific sector during five working days in January 2014. The volume of the water consumed by a given sector can be simply inferred in real-time as an algebraic sum of the flows delivered by its associated flow-meters, in m^3. One can notice a periodicity in the water consumption, related to the human activity.

The paper is organized as follows. Section 2 discusses the CUSUM algorithm and the choice of its parameters, and illustrates the proposed improvements. Section 3 presents the experiments performed to validate the efficiency of the proposed improvements. The paper ends with a conclusion.

2 CUSUM Algorithm

2.1 Algorithm Description

Applied on a data stream, CUSUM takes into account all the past values of the stream by calculating the cumulative sum of the deviations from the target value which is defined as the mean of the observables in the training window, μ_0. It is implicitly assumed that the observed process $(S_t)_{t \geq 0}$ is in-control (stable) during the training window. The cumulative sum control chart C_t, initially set to 0, $(C_0 = 0)$, is calculated as follows:

$$C_t = \sum_{j=1}^{t}(s_j - \mu_0); t \geq 1$$

In order to quantify and detect small variations, CUSUM defines two statistics C_t^+ and C_t^-. C_t^+ accumulates for relatively high values of the observed process, the distance to $(\mu_0 + K)$; K being a given threshold that will be discussed later. For small values of the observed process, the cumulative distance to $(\mu_0 - K)$ is handled by C_t^-.

$$C_t^+ = max[0, s_t - \mu_0 - K + C_{t-1}^+]$$
$$C_t^- = max[0, \mu_0 - s_t - K + C_{t-1}^-]$$

The threshold K is also called the allowance value. It depends on the mean shift that we want to detect. If either C_t^+ or C_t^- exceeds the decision threshold H, the process is considered as out-of-control and an alarm will be signaled. The process is declared as in-control when the cumulative sum is again under the threshold H. K and H are often related to the standard deviation σ_0 calculated in the training window:

$$K = k\sigma_0; \ H = h\sigma_0$$

After the detection of an error, the cumulative sums C_t^+ and C_t^- are reinitialized to 0.

In an improved version of CUSUM called FirCUSUM (for Fast Initial Response) [11], a headstart is introduced to improve the response time of the algorithm. When the process is out-of-control at the start-up, or when it is restarted after an adjustment, the standard CUSUM may be slow in detecting a shift in the process mean that is present immediately after the start-up of the adjustment. To overcome this problem, the headstart consists of setting the starting values C_t^+ and C_t^- equal to some nonzero value, typically $H/2$.

2.2   Choice of the Parameters

The performance of the CUSUM algorithm is closely dependent on the choice of the two key parameters h and k. Two objectives must be achieved when setting these parameters. On the one hand, one must minimize the false positives. In other words, when the process is in-control, ideally, the CUSUM algorithm should not detect any change. On the other hand, any mean shift has to be detected as soon as possible. There is clearly a trade-off between these two objectives. According to [6], it is recommended to take $k = 0.5$ and $h = 4$ or 5. A more complete theoretical study of the performance of CUSUM is provided by Siegmund in [12]. It is based on an approximation of the Average Run Length (ARL) properties.

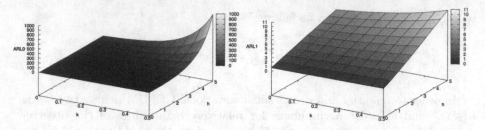

Fig. 2. $ARL_0(k,h)$: impact of k and h on ARL_0

Fig. 3. $ARL_1(k,h)$: impact of k and h on ARL_1

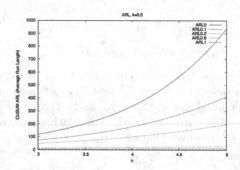

Fig. 4. Impact of the control limit h on ARL_δ

ARL_δ is defined as the expected number of items required by CUSUM to detect a deviation when the process has a mean deviation of δ. The approximation given by [12] is simple compared to other approaches based on approximating transitions from the in-control to the out-of-control state with a Markov chain (see [13]). We choose to focus on positive shifts (the problem is completely analog for negative shifts) so we consider a one-sided CUSUM where only C_i^+ is handled. In this case, Siegmund's approximation of ARL is:

$$ARL = \frac{\exp(-2(\delta - k)h') + 2(\delta - k)h' - 1}{2(\delta - k)^2}$$

where δ is the mean process shift, in the units of σ_0, and $h' = h + 1.166$. In this equation, the observed process is assumed to be normally distributed.

We first plotted in Fig. 2 the variation of ARL_0 based on this equation. Recall that ARL_0 should be high to minimize false positives. According to this Figure, h must be taken at least equal to 3 to have an ARL_0 higher than 100, when k is close to 0.5.

The second step is to minimize ARL_δ to detect the deviation as soon as possible and to achieve a good reactivity. According to Fig. 3, the ARL decreases when h decreases, that is why h should be small to have a better reactivity in case of a mean shift. For $h \in [3, 5]$, ARL_1 is between 5 and 11 which corresponds

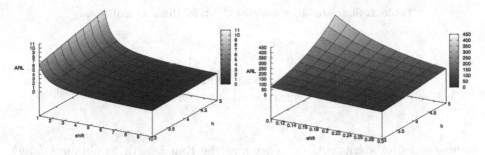

Fig. 5. Impact of the shift δ on ARL, $h \in [3,5]$

Fig. 6. Impact of small shifts δ on ARL, $h \in [3,5]$

to a reasonable response time or reactivity. This result is confirmed by Fig. 5 which shows that CUSUM has a small ARL_δ for large shifts ($\delta \geq 1$).

In Figs. 4 and 6, we focused on very small shifts detection ($\delta \in [0,1]$). When the mean deviation δ is small, ARL_δ becomes high, close to ARL_0. One can conclude that the shift δ has to be at least equal to 0.5 to guarantee an order of magnitude of difference between ARL_0 and ARL_δ ($ARL_0 \sim 100$ and $ARL_\delta \sim 10$).

2.3 Variability of the Run Length

An important criterion used to evaluate the reactivity of a control chart algorithm is the Average Run Length (ARL). We recall that it represents the expected number of observations needed before an out-of-control alarm is detected. ARL has two interpretations. ARL_0 is defined as the average number of in-control data needed by the control chart algorithm to signal a false alarm. ARL_δ is the average number of out-of-control data required to detect the error after a process mean changed. ARL_0 has to be as large as possible, and ARL_δ has to be small in order to detect the shift quickly. The run length is a random variable. To the best of our knowledge, only its average was theoretically studied. No theoretical results about the variability of Run Length are provided in the literature. In this paper, We address this problem using simulations.

We report in Table 1 the average and the standard deviation of the Run Length, respectively ARL and $SDRL$, for different values of shift δ. In order to simulate ARL_0, a sample of size 1000 is generated under an in-control situation. CUSUM chart is then applied to the samples until an out-of-control signal is triggered. The number of observations when the signal is triggered is the in-control Run Length RL_0. We repeat this simulation 150 times so we can get the average value: ARL_0. The considered process follows the standard normal distribution. To simulate the ARL_δ, we performed the same experiments with an injected shift of δ in the process mean. We can notice that the average and the standard deviation of the Run Length decrease with the increase of the shift δ. Moreover, the false positives detected by CUSUM closely depend on the execution, as they are calculated using RL_0 which has a high standard deviation

Table 1. Simulated Run Lengths (RL), for different shift sizes

	Shift δ					
	0	0.25	0.5	1	2	3
Simulated ARL	328	73.59	25.82	8.02	3.38	2.74
SDRL	220	8.38	1.92	1.77	1.41	1.31

compared to its average ARL_0. Therefore the Run Length to obtain a false positive can be significantly lower than ARL_0. One can conclude that for a single execution, the CUSUM can lead quickly to a false positive.

2.4   Enhancing the Reactivity of CUSUM Algorithm

Recall that to detect positive mean shifts, CUSUM is based on the following cumulative statistic:

$$C_t^+ = max[0, s_t - \mu_0 - K + C_{t-1}^+]$$

In the standard version of CUSUM, the deviation from the expected average is declared after performing enough iteration to deeply impact C_t. However, it is not possible to know the exact start time of the small shift. In fact, the change detection occurs after real change start time and no estimation of this latter is provided in the literature. Moreover, the observed process is considered as in-control (end of the deviation) when C_t is less than the detection threshold H. As C_t is a cumulative sum, it sometimes takes a long time (many iterations) to achieve this condition. Thus, the end of deviation is declared a long time after the real return to the standard behavior.

To obtain a complete CUSUM algorithm, we propose the following improvements:

– Add an estimation of the start time of the change.
– Improve the precision of the end time of the change.

Error Start Time. The error real start time estimation for CUSUM algorithm was not addressed before. The change is simply declared when it is detected. Let us take $K = \sigma_0/2$, and s_t a process normally distributed: $S \sim \mathcal{N}(\mu_0, \sigma_0^2)$.

When the process s_t is in-control, $(s_t - \mu_0 - K)$ has the same probability of being positive or negative for symmetry reasons. When s_t has a mean positive shift (of σ_0 as an example), $(s_t - \mu_0 - K)$ becomes very likely to be positive. Therefore, C_t is very likely to be increasing ($C_t > C_{t-1}$). This is the key idea behind our improvement. When the process is out-of-control, the value of C_t is very likely to increase as long as the deviations persist.

When an error is detected ($C_t > H$), the start time of this error can be estimated by the moment where C_t became strictly increasing. This moment can be inferred even if it is former to the detection of the error. For this purpose, we introduce a counter N that we update each time we calculate C_t. If an error is declared at time t, its start time is estimated as: error start time $= t - N + 1$.

Initially, N is set to 0, then, it is updated as follows:

- $N \leftarrow N + 1$ if the value of C_t increases.
- $N \leftarrow N - 1$ if the value of C_t decreases.
- N is reset to 0 if the end of the error is detected.

In the case of an in-control process, N has a random distribution with a null mean. The counter N is used to estimate the size of each error, in other words, the process out-of-control duration.

Error End Time. The second improvement of CUSUM is related to the end of the error. In the standard version of CUSUM, the end of the error is declared when C_t becomes lower than the threshold H. Being a cumulated sum, C_t needs many steps (or iterations) to attain its normal values ($<H$) after the end of the error. To improve the reactivity of the CUSUM algorithm, we introduce a counter Z to be able to detect the end of the error quickly.

The key idea of this improvement is that when C_t becomes constant or decreasing, the current deviation is very likely to be stopped. The condition $C_t^- \leq C_{t-1}^-$ is always achieved before $C_t^- < H$ as in case of error $C_{t-1}^- > H$.

The counter Z is updated each time we calculate C_t. It depicts the number of successive decreases of C_t.

Initially, Z is set to 0, then, it is updated as follows:

- $Z \leftarrow Z + 1$ if C_t decreases.
- $Z \leftarrow 0$ otherwise.

The end of the error is declared when Z exceeds a given threshold Z_0. Just like σ_0 and μ_0, this latter is inferred from the training window. It is the average number of successive decreases in the training window (when the process is in-control).

The detailed pseudo code of the anomaly detection using the improved CUSUM algorithm is presented in Algorithm 1.

3   Experiments and Results

3.1   Efficiency Metrics

In this section, we evaluate the performance of CUSUM algorithm for the change detection. In information retrieval domain, the performance metrics used to evaluate the performance of a change detector are precision, recall, and specificity.

These metrics are based on the True Positives (TP), False Positives (FP), True Negatives (TN) and False Negatives (FN). The errors committed by CUSUM can either be false positives (the process is considered as out-of-control while it is not true) or false negatives (no alarm is signaled whereas the process is out-of-control). *Precision* metric is the proportion of true alarms compared to all the alarms risen by CUSUM. *Recall* metric depicts the true positive rate: the probability that CUSUM algorithm identifies a truly erroneous point. *Specificity* metric represents the true negative rate: the proportion of points considered truly as not erroneous by CUSUM compared to the total number of not erroneous points present in the dataset.

$$Precision = \frac{TP}{TP + FP}; \ Recall = \frac{TP}{TP + FN};$$
$$Specificity = \frac{TN}{TN + FP};$$

Algorithm 1. Improved CUSUM anomaly detection algorithm

Input   : Stream of items $S\ \{s_0, s_1, ..., s_n\}$, $K = \sigma_0/2$, $H = 4\sigma_0$
Output: errorAlarm(), errorStartTime, errorEndTime

1 $C_0 \leftarrow 0$; $N \leftarrow 0$; $Z \leftarrow 0$;
2 *errorInProgress* $\leftarrow$ *false*;

3 **foreach** *incoming item* s_t, $t > 0$ **do**
4     $C_t = max[0,\ s_t - \mu_0 - K + C_{t-1}]$
5   **if** $(C_t > C_{t-1})$ **then**
6        $N++$;
7        $Z = 0$;
8   **if** $(C_t > H)$ **then**
9   *errorAlarm*();
10   *errorStartTime* $= t - N + 1$;
11   *errorInProgress* $= true$;
12   **end**
13   **end**
14   **if** $(C_t < C_{t-1})$ **then**
15       $N--$;
16       $Z++$;
17   **if** $(errorInProgress == true)$ **then**
18   **if** $(Z > Z_0)$ **then**
19   *errorEndTime* $= t - 1$;
20   *errorInProgress* $= false$;
21             $C_t = 0$;
22             $N = 0$;
23   **end**
24   **end**
25   **end**
26 **end**

3.2   Detecting Mean Change

The objective of this section is to validate our proposed improvements of CUSUM algorithm to detect a negative shift of the mean among normally distributed simulated data. Let us take a process $(S_t)_{t \geq 0}$ that follows the standard normal distribution: $S \sim \mathcal{N}(0, 1)$. We first considered 1000 observations of S, then we injected at random moments 10 deviations (also called errors). The purpose of the experiments is to apply CUSUM to detect these deviations and to use our improvements to estimate the start time and the end time of each detected deviation. The errors have a random length taken in $[1, 50]$. The cumulative errors length equals 239 points or observations. As the targeted change is a mean shift, for each error, we replaced the original points by new observations issued from a shifted process $S' \sim \mathcal{N}(-1, 1)$. Notice that the variance of the process remains unchanged.

As only negative shift is considered in this section, we only focus on the cumulative parameter C_t^-, that we simply denote by C in the following parts.

$$C_t^- = max[0, \mu_0 - s_t - K + C_{t-1}^-]$$

k is set to $0, 5$ and the threshold h is taken equal to 4, according to the recommendations given in Sect. 3.

The variation of C over time, together with the detection threshold H are plotted in Fig. 7. We can notice that C is very variable over time and presents several peaks that are mainly caused by the injected errors. CUSUM signals errors each time the value of C exceeds the control limit H. We obtained a total number of 8 alarms corresponding to the 8 detected errors. Two errors are missed because they have very small durations.

Figure 8 depicts the variation of the counter N over time. Recall that N is used to estimate the start time of the error. It is incremented by one with the increase of C and decremented C decreases. Figure 8 shows that when the process is in-control, the value of N has a null average and very small standard deviation. The presence of an error induces a notable increase of N.

The variation of the counter Z over time is shown in Fig. 9. Z counter enables to estimate the end time of the error. It counts the number of successive decreases of C. We can see that Z has small variations with a mean of 0.22 and a standard deviation of 0.57. The end of the error is declared when Z exceeds its average value Z_0 calculated in the training window. In our case, Z_0 equals to 0.25.

The evaluation of the proposed improvements is given in Table 2. The 1000 considered points of the process are first classified into actual error and actual not error. Then we add a second classification according to the detection results of the improved CUSUM. The same experiments are performed with the standard version of CUSUM and FirCUSUM. Recall that FirCUSUM is an improved version of CUSUM. With the classic version of CUSUM algorithm, the values C_t^+ and C_t^- are reset to zero after the detection of a change. The objective of FirCUSUM is to improve the performance of CUSUM by setting the values C_t^+ and C_t^- equal to some nonzero value, typically $H/2$.

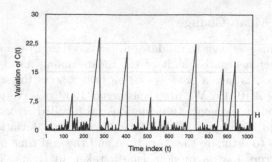

Fig. 7. Variation of C_t over time

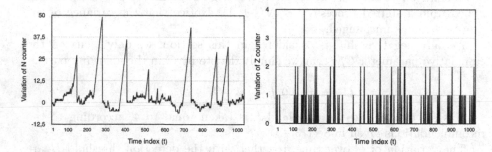

Fig. 8. Variation of N over time   **Fig. 9.** Variation of Z counter over time

Table 2. Obtained results for mean change detection

CUSUM version	Actual	Detected as	
		Error	Not error
Standard	Error	189	50
	Not error	203	558
FirCUSUM	Error	174	65
	Not error	196	565
Improved	Error	**221**	**18**
	Not error	**23**	**738**

The obtained results are presented in Table 2. One can notice that our proposed algorithm outperforms both standard and FirCUSUM. It significantly decreased both false negatives and false positives.

The three efficiency metrics (precision, recall, and specificity) of the standard CUSUM, FirCUSUM and the improved version of CUSUM are given in Table 3. All these metrics are enhanced using our improved version of CUSUM. The proposed improvements give very good results as the three efficiency metrics are above 0.9.

Table 3. Performance metrics of CUSUM

CUSUM version	Precision	Recall	Specificity
Standard	0.48	0.79	0.73
FirCUSUM	0.47	0.72	0.74
Improved	**0.90**	**0.92**	**0.96**

3.3   Application to Stuck-at Error: Detecting Variation Change

In this section, we apply the improved CUSUM algorithm on a real data stream issued from water flow-meters, to detect the so-called stuck-at errors.

During the data analysis process, the conclusions and decisions are based on the data. If the data are dirty, this will leads to defective and faulty results. Improving the data quality is thus inevitable to obtain reliable results. One of the data quality measures is the *accuracy*. It represents the difference between the observation's value and the true value which the sensor aims to represent. Due to the instrumental, physical and human limitations, malfunction and miscalibration of the sensor, the observations values of the sensor can deviate significantly from the true ones. These deviated values are called faults [14].

The stuck-at error, also called CONSTANT in [15,16], is a type of sensor errors. A stuck-at x error occurs when the sensor is stuck on an incorrect value x. The low battery of the sensor, a dead sensor, or the malfunction of the sensor, may cause this error. During such a situation, a set of successive observations will have the same value $x \pm \epsilon$. The error may last a long time, and the sensor may or not return to its normal behavior. [16] showed that CONSTANT errors are present in the sensors data of the INTEL Lab and in NAMOS data set. This kind of error concerns about 20% of their data. The variance is the characteristic to be modeled in order to detect this type of error, during which, the variance of the data drops significantly.

We use the improved CUSUM to detect the stuck-at errors. As the change concerns the variance of the observed process, we choose to apply CUSUM on the variations v_t of the observed values: $v_t = |s_t - s_{t-1}|$. $(V_t)_{t \geq 0}$ is a positive time series with an average of μ_0 and a standard deviation σ_0 during the training window. As the stuck-at error engenders a decrease of μ_0, we only focus here on the cumulative statistic C_t^-, for a one-sided CUSUM.

The dataset duration considered in this section is of 10 days in January 2014, with a total number of 960 observations. To inject a stuck-at error in the time series, we chose random instant t and we replaced s_t by a random variable uniformly distributed in $[s_t - \mu_0 + \sigma_0, s_t + \mu_0 - \sigma_0]$, for a random number of successive values beginning from the instant t. Hence the mean of the process $v_t = |s_t - s_{t-1}|$ drops from μ_0 to $\mu_0 - \sigma_0$. The shift that we want to detect is about $-\sigma_0$. We repeated this mechanism 10 times to inject 10 errors. The errors have a random length taken in $[1, 50]$. We obtained a total number of 263 injected erroneous points.

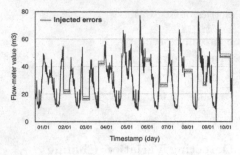

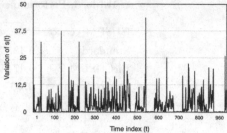

Fig. 10. Injection of variation changes

Fig. 11. Variation v_t of s_t after the injection of errors

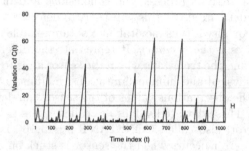

Fig. 12. Variation of C_t over time

Figure 10 shows the considered dataset s_t with the injected errors. The variations of s_t denoted as $v_t = |s_t - s_{t-1}|$ are plotted in Fig. 11. One can easily notice a periodicity in the water consumption with a difference between the working and not working days. Moreover, during the injected errors, the variations of s_t denoted as $v_t = |s_t - s_{t-1}|$ drop significantly. Recall that we performed CUSUM on v_t.

Figure 12 displays the variation of C_t over time. 8 alarms were signaled by CUSUM, when C_t exceeds the threshold H. They correspond to real injected errors. CUSUM missed two errors as they have very small durations (of only 1 and 4), compared to ARL_1. In fact, according to Sect. 3.2, ARL_1 equals 8.38 (when $k = 0.5$, $\delta = \sigma$ and $h = 4$). It means that we need in average 8 observations to could detect this change.

We recap in Tables 4 and 5 the obtained results of the variation change detection, after applying the improved and the standard version of CUSUM algorithm. Just like in the previous subsection (using simulation), we checked, based on these values that the proposed improvements enhance the three efficiency metrics: the precision, the recall and the specificity of the CUSUM algorithm (see Table 5). Moreover, these results show that the CUSUM algorithm performs good results even if the considered data set does not verify the normal distribution.

Table 4. Obtained results for stuck-at errors detection

CUSUM version	Actual	Detected as	
		Stuck-at error	Not stuck-at error
Standard	Stuck-at error	134	129
	Not stuck-at error	121	576
Improved	Stuck-at error	**255**	**8**
	Not stuck-at error	**68**	**629**

Table 5. Performance metrics of CUSUM

CUSUM version	Precision	Recall	Specificity
Standard	0.52	0.50	0.82
Improved	**0.78**	**0.97**	**0.90**

4 Conclusion

We presented in this paper an in-depth study of CUSUM control chart algorithm. We first analyzed the average and the variations of the Run Length, which qualifies both false positives and the response time of the algorithm in case of a change detection. Then, we introduced an improvement of the precision of the estimation of the start time and the end time of each detected change. The designed algorithm performs good results against both synthetic and real datasets. We also adapted the CUSUM to deal with the stuck-at error where the variability of the process is addressed.

References

1. Chabchoub, Y., Chiky, R., Dogan, B.: How can sliding HyperLogLog and EWMA detect port scan attacks in IP traffic? EURASIP J. Inf. Secur. **2014**(1), 5 (2014)
2. Manonmani, R., Mary Divya Suganya, G.: Remote sensing and GIS application in change detection study in urban zone using multi temporal satellite. Int. J. Geomat. Geosci. **1**(1), 60 (2010)
3. Basseville, M., Nikiforov, I.V., et al.: Detection of Abrupt Changes: Theory and Application, vol. 104. Prentice Hall, Englewood Cliffs (1993)
4. Roberts, S.W.: Control chart tests based on geometric moving averages. Technometrics **1**(3), 239–250 (1959)
5. Page, E.S.: Continuous inspection schemes. Biometrika **41**(1/2), 100–115 (1954)
6. Montgomery, D.C.: Introduction to Statistical Quality Control. Wiley, New York (2007)
7. Ewan, W.D.: When and how to use CUSUM charts. Technometrics **5**(1), 1–22 (1963)
8. Bissell, A.F.: CUSUM techniques for quality control. Appl. Stat. **18**, 1–30 (1969)
9. Goel, A.L., Wu, S.M.: Economically optimum design of CUSUM charts. Manag. Sci. **19**(11), 1271–1282 (1973)

10. Reynolds, M.R.: Approximations to the average run length in cumulative sum control charts. Technometrics **17**(1), 65–71 (1975)
11. Lucas, J.M., Crosier, R.B.: Fast initial response for CUSUM quality-control schemes: give your CUSUM a head start. Technometrics **24**(3), 199–205 (1982)
12. Siegmund, D.: Sequential Analysis: Tests and Confidence Intervals. Springer, New York (2013). https://doi.org/10.1007/978-1-4757-1862-1
13. Brook, D., Evans, D.A.: An approach to the probability distribution of CUSUM run length. Biometrika **59**, 539–549 (1972)
14. El Sibai, R., Chabchoub, Y., Chiky, R., Demerjian, J., Barbar, K.: Assessing and improving sensors data quality in streaming context. In: Nguyen, N.T., Papadopoulos, G.A., Jędrzejowicz, P., Trawiński, B., Vossen, G. (eds.) ICCCI 2017. LNCS (LNAI), vol. 10449, pp. 590–599. Springer, Cham (2017). https://doi.org/10.1007/978-3-319-67077-5_57
15. Ramanathan, N., Schoellhammer, T., Estrin, D., Hansen, M., Harmon, T., Kohler, E., Srivastava, M.: The final frontier: embedding networked sensors in the soil. Center for Embedded Network Sensing (2006)
16. Sharma, A.B., Golubchik, L., Govindan, R.: Sensor faults: detection methods and prevalence in real-world datasets. ACM Trans. Sens. Netw. (TOSN) **6**(3), 23 (2010)

Application of the KDD Process for the Visualization of Integrated Geo-Referenced Textual Data from the Pre-processing Phase

Flavio Gomez[1(✉)], Diego Iquira[1], and Ana Maria Cuadros[2]

[1] National University of San Agustín, Arequipa, Peru
fgomeza@unsa.edu.pe, diego452270@gmail.com
[2] La Salle University, Arequipa, Peru
anamariacuadros@gmail.com

Abstract. Geo-referenced textual data has been the subject of multiple investigations, by providing opportunities to better understand certain phenomena according to the content that is shared, either on-line such as social networks, blogs, and news; or through repositories such as scientific research articles, geo-referenced virtual books, among others. However, the characteristics of this information are studied, analyzed and processed separately, either through its textual components or its geo-spatial components, which offers a separate understanding of the results.

In this paper, we propose an integration of textual and geo-spatial components from the pre-processing phase to the visualization stage, As a part of the Document Mapping process based on the phases of the Knowledge Discovery in Databases (KDD). Achieving two main results (1) minimize the problems that arise in the visual phase, such as data occlusion and (2) provide a more detailed understanding between the textual relationships of the data when plotted in a geo-spatial map.

Keywords: GIS · Textual data · Geo-referenced data · KDD
Document map · Data visualization · Integration

1 Introduction

First of all, the amount of complex data that is continuously generated on the web has reached a scale without precedents, additionally this data present textual content and geo-spatial information together, for this reason it is known as geo-referenced textual data or geo-textual data.

Meanwhile, the spatial dimension as a complement to the textual content has allowed adding a new semantically rich aspect for the analysis [3]. However, when the visualization of all this information is very large and exceeds the capacity of a single view by the user, the user has difficulty in compressing the data [9].

As a result of the high dimension of the textual data and the need to explore the relationship understand them, so the use of document maps based on content

© Springer International Publishing AG, part of Springer Nature 2018
M. R. Luaces and F. Karimipour (Eds.): W2GIS 2018, LNCS 10819, pp. 41–50, 2018.
https://doi.org/10.1007/978-3-319-90053-7_5

is proposed, in order to use the projection techniques to solve the problem of multidimensionality. One of the main challenges that arise when working with geo-referenced textual content occurs in the pre-processing phase, specifically during the construction of the vector of characteristics [1], because the spatial and textual characteristics come intrinsically together, thus conditioning the vector of characteristics so that it represents the textual and geo-spatial content in an integrated manner, which generates a correct association between the geographical coordinates and the textual content of the data [19].

On the other hand, the extraction of information in geo-referenced textual data implies the development of different steps or phases, which can be carried out as part of the process of the Knowledge Discovery in Databases (KDD) [18]. For example, data mining offers solutions to find and discover patterns in large amounts of Geo-referenced textual data, which prior to this process are unknown [8], But despite being a fundamental step for the KDD process [13], the results it offers remain complex and difficult for the user to understand. To sum up the visualization of data, the visual exploration and the interaction with the user play an important role in the decision making [4], particularly if working with large multidimensional data sets that need to be explored and understood.

This paper is organized as follow: Sect. 2 presents the concepts for the development of the proposal, specifically about KDD process based on document maps, Sect. 3 shows the proposed method, Sect. 4 shows the visualization results. Finally, in Sect. 5, presents the conclusions of the paper.

2   Background

First, the geo-referenced textual data has been the subject of multiple investigations [8], providing opportunities to better understand certain phenomena according to the information that is shared online [5]. For example, frequency calculation of terms for the development of recommendation systems based on geo-referenced content shared in social networks [2,19], the use of density-based algorithms to identify and predict escape routes in real time when natural disasters occur according to data shared on the web [7], sentiment analysis of geo-textual content applied to keywords [19], the application of statistical models to compare the population of twitter users with the demographics of a country [12], the analysis of spatial distribution on social content generated during presidential elections in the United States [17], the correlation analysis between meteorological data and shared keywords on twitter [11], the use of probabilistic models to locate the epicenter of seismic movements by classifying keywords in tweets [16], among others.

To summarize, there are different types of methods to process geo-referenced textual data, where the KDD process based on document maps has been chosen to be used in this paper and which is explained below.

2.1   KDD Process Based on Document Maps

Mohamed [13] defines the three main phases that form the KDD process as:

- The data preparation phase that consists in selecting, cleaning and transforming data in a compatible format for the next phase.
- The data mining phase that is the central axis of the KDD process, which allows extracting relevant and interesting patterns (not trivial, implicit, previously unknown and potentially useful) from large amounts of data through the application of intelligent methods.
- The evaluation and interpretation phase during which the generated patterns are interpreted and evaluated for the integration of knowledge in the decision-making stage.

However in the field of the KDD process, the capacity to generate data greatly exceeds the ability to interpret the information [10]. For this reason, a total compression of the user's results is difficult [10], so in order to explore the relationships of geo-referenced textual data, the use of document maps based on content is proposed, in order to use the projection techniques to solve the problem of the multidimensionality of the data, whose general process consists of three main steps: (1) Pre-processing; (2) Projection and triangulation; and (3) Mapping of additional information [15], these steps can be appreciated in Fig. 1.

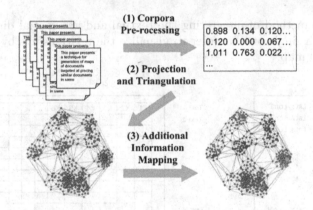

Fig. 1. Overview of the document-based document mapping process [15]

Document maps are visual information spaces that allow user navigation; in the map, the relationships of similarity between data are reflected, by the use of different geometric representations.

The first step in this process is the pre-processing phase where the documents in the collection are converted into a characteristic vector in a multidimensional space based on the number of terms. Then, the dimensionality of these vectors is reduced to only two dimensions to represent each document as a point on a Cartesian plane. Once these points are projected on a plane, they are triangulated, to reflect the proximity between them and also to generate a view of how these documents are related. Proximity or measure of similarity is a metric that measures the closeness between objects (similarity) or the distance between objects (dissimilarity). Both metrics are complementary, meaning that the greater the index of similarity between two objects, the lower their dissimilarity index. The distance most used to perform this calculation is the cosine distance, which is not properly a distance but a measure of similarity between two vectors in a space that has an interior product defined. In the Euclidean space, this inner product is the scalar product represented by:

$$\vec{x_1} \cdot \vec{x_2} = \|x_1\| \, \|x_2\| \tag{1}$$

And the measure of similarity is given by:

$$\cos(\theta) = \frac{\vec{x_1} \cdot \vec{x_2}}{\|x_1\| \, \|x_2\|} \tag{2}$$

Finally, it is possible to map more information by choosing different visual attributes for each type of document such as the color or size of each point, thus generating clusters, which provides additional information about the documents.

3   Proposed Method

We propose a method of integrating the textual and geo-spatial dimensions of the data into a single one, in order to visualize the relationships by content on a geographical map.

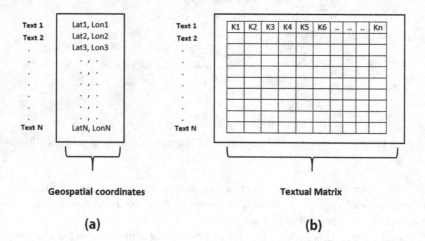

Fig. 2. Representation of the feature vectors. (a) Geo-spatial vector conformed by geographical coordinates: latitude and longitude, and (b) Textual vector conformed by a matrix of similarity between the data whose distance was obtained by cosine distance, to determine the similarity of the content.

3.1   Data and Feature Vector

The data that was used on the proposal is a collection of abstracts of scientific articles of different topics, which were published in the United States of America, where each article was geo-referenced using the coordinates (latitude and longitude) of the university to which belongs to the principal author of each one.

The stored data that was obtained was converted into a feature vector. Then the vector of geo-spatial characteristics was determined based on the geographic coordinates, latitude and longitude, Fig. 2(a), while the textual vector was determined by a matrix of similarity between the points, whose distance was obtained by the cosine distance, Fig. 2(b), where the value oscillates between 0 (completely different) and 1 (completely similar).

However, due to the geo-referenced textual data, and having as purpose that the dimensions of content and spatiality are jointly and integrated from the pre-processing phase, the characteristic vector is determined by a resulting matrix.

The resulting matrix was obtained by adding the Euclidean distance matrices (for spatiality) and cosine distances (for content). In addition, the sum of both matrices is determined according to a proportion or weight for each of them as show in Eq. 3.

$$M_r = \frac{\beta(M_t) + \alpha(M_e)}{2} \tag{3}$$

where the resulting matrix M_r is equal to the mean between the sum of the textual matrix M_t multiplied by a constant β and the spatial matrix M_e multiplied by a constant α, besides $\beta + \alpha = 1$.

The alpha and beta values are established according to the matrix that is desired to have the greatest weight in the visualization. In this way, if beta is greater than alpha, the matrix that will have the greatest weight will be the textual, in the opposite case, if beta in less than alpha, the matrix with the greatest weight will be the spatial one.

4   Result of Clustering and Visualization

Before working with the resulting matrix, the neighbor joining algorithm was used separately from the text and geo-spatial matrices, which provided similarity relationships by content between the different data points, as shown in Fig. 3 (Left) where four groups are distinguished and whose nodes are close to each other, which represent the 4 main themes of the content. However, when the neighbor joining algorithm is applied to the same data set, but this time according to its spatial distance matrix, the relation of one to another is lost and they are located in a dispersed manner without any apparent relationship as can be seen in Fig. 3 (right).

Based on the Cartesian plane where the relationships obtained by the neighbor joining are shown, it is determined that they are in a better way when processed by their textual content, and distorted when processed geo-spatially.

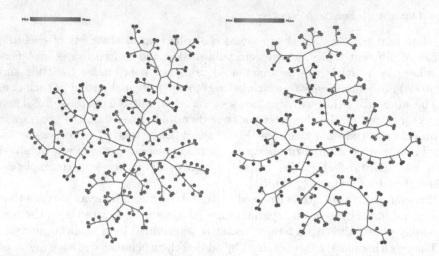

Fig. 3. Clusters representation, where each color represents a specific theme, the textual relations are located on the left, made according to the content of each article, where articles with similar content are located closer. On the other hand, spatial relationships are located on the right, where articles with close coordinates are located closer. Both representations were made using the Neighbor Joining algorithm. Display generated using the PEX tool. (Color figure online)

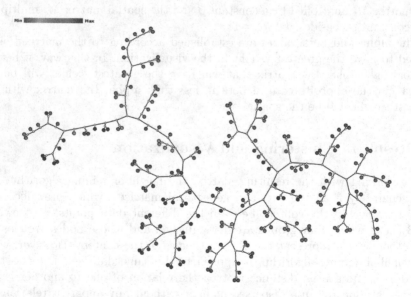

Fig. 4. Visual representation of the resulting matrix between the spatial and textual matrix to be weighted by the equation of Sect. 3.1 where the values of alpha and beta are both 0.5, creating an average proportion between the textual and spatial content, which was generated using the PEX tool [14]. (Color figure online)

In order to find a middle point between both visualizations, formula 3 was used to unite the pre-processing of both matrices and obtain an integrated visualization, as shown in Fig. 4, where the 4 large groups (represented by the colors red, blue, green and yellow) have been separated as sets, but still maintaining the relationship between nodes.

The base of the geo-spatial visualization is carried out using cartograms, since these preserve the traditional geographic maps, maintaining similar forms and conserving the adjacencies between the represented areas. All this, makes it a very useful graphical variable when visualizing heterogeneous data, having great influence on what the user perceives on a map [6].

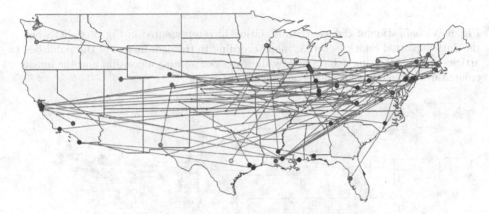

Fig. 5. Visualization of the textual relationships represented in Fig. 3 (left) when the data is plotted on a geographical map according to the coordinates of the published articles, the result being determined by the algorithm *Neighbor Joining* and the image created using the QGis visualization tool.

When the relations between nodes found by the neighbor joining were taken from a Cartesian plane to a geographical plane, the proximity between nodes that was given by similarity of content was lost, due to the different locations of coordinates that each point has, generating from this way occlusion and a cross between the relations, as seen in Fig. 5, whereas when the spatial relationship is visualized the crossing between the relationships of the nodes is non-existent, Fig. 6.

Taking into account that what we seek to achieve is an intermediate and integrated visualization between the textual and geo-spatial dimensions without losing the relationship between the nodes, Fig. 7 shows such integration, with a considerable reduction of the crossings of the edges in comparison with the textual relationship, but without losing relationships or similarities by content.

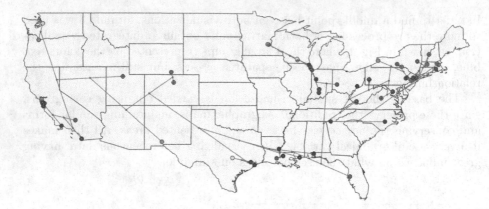

Fig. 6. Visualization of the geospatial relationships represented in Fig. 3 (right) when the data is plotted on a geographic map according to the coordinates of the published articles, the result being determined by the *Neighbor Joining* algorithm and the image created using the QGis visualization tool.

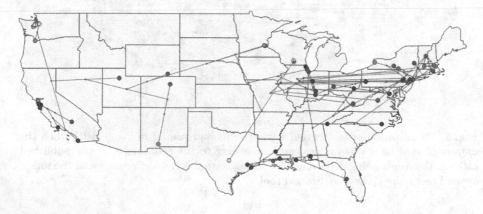

Fig. 7. Combined display of the textual relationships represented in Fig. 4, when plotted on a geographical map, the result being determined by the Neighbor Joining algorithm and the image created using the QGis visualization tool.

5   Conclusions

- It was concluded that there is a strong relationship between the textual and geo-spatial characteristics of the data at the moment of performing analysis and interpretation of the results by the user, which provide valuable information that could not be acquired by working them separately.
- It is possible to achieve a visual improvement and reduce the problem of occlusion and overlapping of the data using only an integrated matrix from the pre-processing phase, increasing in this way the possibilities of improving the understanding and analysis by the user if it is added some contribution to the phase of data visualization.

- The method that we propose is very useful to visualize textual data without losing its relationships by content when these are plotted on a geographical map. For example, the proposed method can be applied in various cases such as: News forums to understand what relationships exist between certain news items according to the topic and the place where the events occurred, another example is to determine the relationship between the content and the location of publications on Twitter, among other cases.
- Finally, the visualization in the geo-spatial map can be improved using 3D visualization tools to separate each cluster by layers.

Acknowledgements. The present work was achieved thanks to the joint work with my advisor, for her persistence and tenacity at the moment of sharing her teachings with me, to my distinguished teachers who have forged knowledge from the first day of classes, whom with nobility and enthusiasm influenced as an example in me and my colleagues in the master's degree in computer science; also thanks to CONCYTEC, FONDECYT and Cienciactiva for the support and opportunities provided that made this work possible.

References

1. Aggarwal, C.C., Zhai, C.: A survey of text clustering algorithms. In: Aggarwal, C., Zhai, C. (eds.) Mining Text Data, pp. 77–128. Springer, Boston (2012). https://doi.org/10.1007/978-1-4614-3223-4_4

2. Bao, J., Zheng, Y., Mokbel, M.F.: Location-based and preference-aware recommendation using sparse geo-social networking data. In: Proceedings of the 20th International Conference on Advances in Geographic Information Systems, pp. 199–208. ACM (2012)

3. Cong, G., Feng, K., Zhao, K.: Querying and mining geo-textual data for exploration: challenges and opportunities. In: IEEE 32nd International Conference on Data Engineering Workshops (ICDEW), 2016, pp. 165–168. IEEE (2016)

4. De Oliveira, M.C.F., Levkowitz, H.: From visual data exploration to visual data mining: a survey. IEEE Trans. Vis. Comput. Graph. **9**(3), 378–394 (2003)

5. Doytsher, Y., Galon, B., Kanza, Y.: Querying geo-social data by bridging spatial networks and social networks. In: Proceedings of the 2nd ACM SIGSPATIAL International Workshop on Location Based Social Networks, pp. 39–46. ACM (2010)

6. Dykes, J., MacEachren, A.M., Kraak, M.J.: Moving geovisualization toward support for group work. In: Exploring Geovisualization, p. 445 (2005)

7. Jain, S.: Real-Time Social Network Data Mining for Predicting the Path for a Disaster (2015)

8. Lee, C.-H., Yang, H.-C., Wang, S.-H.: A location based text mining approach for geospatial data mining. In: Fourth International Conference on Innovative Computing, Information and Control (ICICIC), 2009, pp. 1172–1175. IEEE (2009)

9. Liu, S., Cui, W., Yingcai, W., Liu, M.: A survey on information visualization: recent advances and challenges. Vis. Comput. **30**(12), 1373–1393 (2014)

10. Ltifi, H., Ayed, M.B., Alimi, A.M., Lepreux, S.: Survey of information visualization techniques for exploitation in KDD. In: IEEE/ACS International Conference on Computer Systems and Applications, 2009, AICCSA 2009, pp. 218–225. IEEE (2009)

11. Lwin, K.K., Zettsu, K., Sugiura, K.: Geovisualization and correlation analysis between geotagged Twitter and JMA rainfall data: case of heavy rain disaster in Hiroshima. In: 2015 2nd IEEE International Conference on Spatial Data Mining and Geographical Knowledge Services (ICSDM), pp. 71–76. IEEE (2015)
12. Mislove, A., Lehmann, S., Ahn, Y.-Y., Onnela, J.-P., Rosenquist, J.N.: Understanding the demographics of twitter users. In: ICWSM, 11:5th (2011)
13. Mohamed, E.B., Ltifi, H., Ayed, M.B.: Integration of temporal data visualization techniques in a KDD-based DSS application in the medical field. J. Netw. Innovative Comput. **2**, 061–070 (2014)
14. Paulovich, F.V., Oliveira, M.C.F., Minghim, R.: The projection explorer: a flexible tool for projection-based multidimensional visualization. In: XX Brazilian Symposium on Computer Graphics and Image Processing, 2007, SIBGRAPI 2007, pp. 27–36. IEEE (2007)
15. Paulovich, F.V., Minghim, R.: Text map explorer: a tool to create and explore document maps. In: Tenth International Conference on Information Visualisation (IV 2006), pp. 245–251. IEEE (2006)
16. Sakaki, T., Okazaki, M., Matsuo, Y.: Earthquake shakes Twitter users: real-time event detection by social sensors. In: Proceedings of the 19th International Conference on World Wide Web, pp. 851–860. ACM (2010)
17. Tsou, M.-H., Yang, J.-A., Lusher, D., Han, S., Spitzberg, B., Gawron, J.M., Gupta, D., An, L.: Mapping social activities and concepts with social media (Twitter) and web search engines (Yahoo and Bing): a case study in 2012 US presidential election. Cartography Geogr. Inf. Sci. **40**(4), 337–348 (2013)
18. Wüthrich, B.: Knowledge Discovery in Databases (1995)
19. Zhao, K., Liu, Y., Yuan, Q., Chen, L., Chen, Z., Cong, G.: Towards personalized maps: mining user preferences from geo-textual data. Proc. VLDB Endow. **9**(13), 1545–1548 (2016)

GeoSPRINGS: Towards
a Location-Aware Mobile Agent Platform

Sergio Ilarri[1]([✉]), Pedro Roig[2], and Raquel Trillo[1]

[1] Department of Computer Science and Systems Engineering,
University of Zaragoza, I3A, Zaragoza, Spain
{silarri,raqueltl}@unizar.es
[2] Department of Computer Science and Systems Engineering,
University of Zaragoza, Zaragoza, Spain
peter.roig@gmail.com

Abstract. Mobile agent technology enables the autonomous migration
of code from one computer/device to another, to efficiently exploit the
available computing resources or access data locally. It can provide interesting advantages in distributed and mobile computing scenarios. However, existing mobile agent platforms do not offer facilities for the design
of agents that are location-aware and that can support a variety of communication options, including mobile ad hoc communications.

In this paper, we present our ongoing work towards the development
of a location-aware mobile agent platform, that we call GeoSPRINGS.
The platform provides functionalities for agents to travel to specific geographic locations/areas and to perform geo-cast communications, as well
as the capability to operate using both wide-area cellular networks and
direct ad hoc communications among the mobile devices.

Keywords: Mobile agents · Location-awareness · Mobile devices
Android

1 Introduction

Mobile agents are pieces of code that have the capability to autonomously
pause their execution on the device/computer where they are executing, move to
another computer or device, and resume their execution at the destination [10].
They rely on the availability of a mobile agent platform [15], that holds one
or several execution environments where the mobile agents execute. A mobile
agent platform usually provides services related to the agent execution, lifecy-
cle management, agent communication by using Remote Procedure Calls (RPC)
and/or message passing, movements from one execution environment to another,
security [4], and persistency/storage facilities.

Mobile agent technology can provide a wide range of advantages [9], particularly in distributed and mobile computing environments (e.g., see [7,13,16,19]).
Among the frequently-cited benefits of mobile agents, we can mention their

© Springer International Publishing AG, part of Springer Nature 2018
M. R. Luaces and F. Karimipour (Eds.): W2GIS 2018, LNCS 10819, pp. 51–60, 2018.
https://doi.org/10.1007/978-3-319-90053-7_6

autonomy, flexibility, and the effective usage of network bandwidth that they can provide. Thus, for example, a mobile agent can decide to move to another computer or device to process data stored locally there; in such a way, it can collect and transmit through the network only the data that are really required (i.e., unnecessary data are filtered at the data source, avoiding unneeded communications). Another advantage is that they support asynchronous users (i.e., users that are not necessarily connected all the time), which is a very interesting feature for mobile computing; for example, a user could launch a mobile agent to execute a task in another computer/device and, whereas the mobile agent executes that task (possibly moving through other computers or devices, as needed), he/she can disconnect from the network and even turn his/her mobile device off.

Although mobile agent technology appeared in the 1990s, it is still a subject of research. Recent works proposing its use include [16,19] and a recent survey on security aspects can be found in [4]. Besides, some concepts and ideas behind mobile agents are similar to those found in newer topics that can be considered to be still hot research areas, such as *cyber foraging*, where *"mobile devices offload computation to servers called surrogates, which are located in close proximity to the mobile device"* [5]. Mobile agents could be useful in the context of *edge and fog computing* [11,18], for example to implement appropriate techniques to offload computations [5,12,14]. Incorporating geographic migration and communication capabilities in a mobile agent platform could facilitate the development of applications that require monitoring (or acting upon) a geographic area, in scenarios such as environment monitoring, surveillance, emergency notification, and data management in *vehicular ad hoc networks (VANETs)* [16], etc. However, as far as we know, there is no mobile agent platform with location-awareness and with the needed functionalities to operate in mobile ad hoc networks.

2   Development of GeoSPRINGS

In this section, we discuss the main features of GeoSPRINGS and some key design decisions taken during its development.

2.1   From SPRINGS to GeoSPRINGS

GeoSPRINGS is a redesign of *SPRINGS (Scalable Platform foR movING Software)* [8]. SPRINGS is a mobile agent platform focused on achieving scalability and it has shown a high reliability and performance as compared to other approaches (JADE, Aglets, Voyager, Tryllian, and Grasshopper) [15], which motivates our decision to use it as a basis for this work. The architecture of SPRINGS is composed by two basic elements:

– A *SPRINGS context* is an execution environment for the mobile agents. It is basically a Java process with a remote object, called *ContextManager*, listening on a certain communication port. It offers functionalities such as location transparency for agent communications and agent migrations.

- A *SPRINGS region* is a set of contexts within the same administrative domain. A remote object, called *Region Name Server (RNS)*, keeps track of information about its region and performs management tasks (e.g., to guarantee that agent names and context names are unique in the region, translate context names to context addresses, and allocate agent tracking tasks to certain SPRINGS contexts).

As compared to other mobile agent platforms, SPRINGS offers interesting functionalities, such as:

- It provides *location transparency*, which means that agents need neither to handle network addresses when communicating with other agents or when traveling to other SPRINGS contexts nor to explicitly retrieve their locations first. Instead, an agent can communicate with another one by simply specifying the name of the target agent; similarly, an agent can move to another SPRINGS context by just indicating the target context's name. The location transparency for agent calls is achieved thanks to the concept of *dynamic proxy*. When an agent wants to communicate with another one, it first obtains a proxy to that target agent, which is a reference that points to the agent and is automatically updated when it moves to another location.
- It achieves a high *scalability*, as it supports the simultaneous execution of a large number of mobile agents.
- It deals with the *livelock* problem that may occur if an agent moves quickly from one execution environment to another and, due to this, a pending communication with that agent can never get through (when the proxy has been updated, the mobile agent has already moved to another execution environment, and so on).

SPRINGS has been designed to work on distributed computers such as laptops or nodes in a fixed network or mobile devices with a standard full-fledged Java Virtual Machine (JVM), and it is based on the use of Java Remote Method Invocation (RMI). On the other hand, GeoSPRINGS focuses on the execution of mobile agents on Android mobile devices, where RMI is not supported, it enables different communication options (3G, WiFi, WiFi Direct, etc.), and it offers geographic routing capabilities inside the platform itself.

2.2 Design Based on Sockets and a Proxy Pattern

Although a preliminary port of SPRINGS for Android based on LipeRMI was also developed previously, in GeoSPRINGS we decided not to rely on external libraries such as LipeRMI [1] or RipeRMI [6]. So, inspired by RMI, we designed a solution based on sockets and a proxy pattern. To differentiate it from the concept of dynamic proxy presented in Sect. 2.1, in our design we use the term *wrapper* rather than *proxy*, even though using proxy (placeholder for another object) would be more suitable. We see a proxy as a wrapper that encapsulates access to a remote object, thus enhancing the object with remote access.

On the one hand, there is a *WrapperServer* class that opens a socket to listen to remote calls and creates a thread (instance of a *ClientThread* class) every time it receives a connection from a client object, in order to process the call asynchronously. On the other hand, there is a *Wrapper* class (implementing the *java.lang.reflect.InvocationHandler* interface) that encapsulates the access to a remote object and provides a static method *createWrapper* to create a wrapper around any object that needs to be contacted remotely. In this way, the wrapper hides the details needed to perform the remote communications: when a call to one of the methods of the wrapper takes place, a socket is used to open a connection with the real remote object, a request is sent to that object through that socket, and then the answer is read back from the socket. So, the client object interacts locally with the wrapper and all the remote communications are performed transparently. The data sent through the socket as a request are interpreted as the data needed to perform a remote call (name of the target object, name of the method to call, arguments, etc.) and the data sent back from the remote object are the results of the call or potential exceptions thrown during the execution of the method called. To create wrappers and enable a transparent communication with the remote objects, we make use of the *Java Reflection API* and the *Proxy* class. As an example, we show the *createWrapper* method of the *Wrapper* class:

```
public static <T> T createWrapper(String host, int port, String
    serviceName, Class<T> interfaceClass) {
    Wrapper remote = new Wrapper();
    remote._interfaceClass = interfaceClass;
    remote._host = host; remote._port = port;
    remote._serviceName = serviceName; /* Name of the entity (or
        service) registered as a remote object. */
    return (T)
        Proxy.newProxyInstance(interfaceClass.getClassLoader(), new
        Class<?>[]{interfaceClass}, remote);
}
```

2.3 Support for Geographic Locations

As opposed to SPRINGS, GeoSPRINGS is location-aware. For that purpose, we have defined a *GPSLocation class* that is attached to every *ContextAddress* (class that stores data used internally by the platform when a GeoSPRINGS context needs to be contacted, such as its name, hostname, and communication port). A GPSLocation stores the latitude and longitude of the GeoSPRINGS context and provides some utility methods to manage geographic locations (e.g., to check if the location is within a specific circular or rectangular area).

When a GeoSPRINGS context is created on a fixed computer, its GPS location can be provided as a parameter of the creation method. For the case of GeoSPRINGS contexts created on a mobile device (Android smartphone or tablet), the GPS receiver of the device can be dynamically queried to determine the geographic location of the GeoSPRINGS context, by using the *Google*

Play services location APIs [3]. The retrieved information can be queried periodically in order to dynamically update the location of the GeoSPRINGS context with the required refreshment frequency. When the desired refreshment period expires, a GeoSPRINGS context updates its geographic location by calling a method of its corresponding RNS (in case it is connected to an RNS at that moment). In this way, GeoSPRINGS contexts within the region can query the location of the other GeoSPRINGS contexts through the corresponding RNS.

2.4   Location-Aware Movements and Communications

We defined two location-based migration methods in the *SpringsAgent* interface, which allow a mobile agent to travel to a specific circular or rectangular area:

```
public void moveToLocation(GPSLocation loc, double radius) throws
    AgentMovementException
public void moveToLocation(GPSLocation lowerLeftCorner, GPSLocation
    topRightCorner) throws AgentMovementException
```

For the implementation of those methods, we currently use a simple greedy strategy that tries to minimize the distance to the destination: from all the GeoSPRINGS contexts known by the current GeoSPRINGS context, the agent tries to iteratively jump to the one that is geographically closer to the intended destination, and so on. Nevertheless, alternative hopping strategies, using other heuristics or exploiting additional information that may be available (e.g., using information about the expected trajectories of the mobile devices, considering road maps for mobile devices located inside vehicles, etc.), could be considered and integrated [16].

Similarly, we have defined two methods to enable geo-cast communications among agents based on geographic criteria (i.e., call a method on all the mobile agents located within a given circular or rectangular area and obtain the results):

```
public ArrayList<Object> callAgentMethod(GPSLocation center, double
    radius, String methodName, Object[] args)
public ArrayList<Object> callAgentMethod(GPSLocation lowerLeftCorner,
    GPSLocation topRightCorner, String methodName, Object[] args)
```

2.5   Ad Hoc Communications with WiFi Direct

Another design goal for GeoSPRINGS was to enable direct ad hoc communications among mobile devices, even if they do not have Internet connection. For that purpose, we use WiFi Direct [17], which allows establishing a WiFi connection between two devices without the need of an intermediate WiFi access point; besides, any device can be simultaneously connected to another WiFi network.

WiFi Direct allows obtaining the MAC addresses of the nearby devices, rather than their IP addresses. Therefore, to enable the communication with those devices, a previous step to establish a connection is needed. Afterwards, the connection can be used to transmit agents to those GeoSPRINGS contexts or call methods of agents executing there. To implement ad hoc communications in GeoSPRINGS, the *wifi.p2p* library, belonging to the Android framework, was used. A minor nuisance with WiFi Direct is that the first connection between two devices through WiFi Direct requires an explicit authorization by the user, who will need to accept the connection invitation shown in a dialog window.

It should be noted that GeoSPRINGS is able to operate in one of two modes: with Internet connection or with ad hoc connection. When a mobile device has *Internet connection*, then it has access to the RNS and all the functionalities of the mobile agent platform are available. However, when operating in *ad hoc mode*, access to an RNS is not assumed and the functionalities available are more limited; indeed, only the nearby GeoSPRINGS contexts reachable through WiFi Direct can be considered in this case. So, if a mobile agent arrives in a GeoSPRINGS context with no connection to an RNS, then the mobile agent will not be able to use any functionality that requires a connection to an RNS, such as those related to the use of dynamic proxies or performing movements to other GeoSPRINGS contexts by just specifying the context name. In particular, the following methods, already present in the SPRINGS API, can be executed only when there is access to an RNS:

```
public void moveTo(String contextName) throws AgentMovementException
public void moveToURL(String url) throws AgentMovementException
public Object callAgentMethod(String targetAgent, String methodName,
    Object[] args) throws MethodCallException, CommunicationException
```

For example, if an agent executes *moveTo(String contextName)* when there is no connection to an RNS, then an exception will be thrown. Similarly, dynamic proxies pointing to mobile agents executing in GeoSPRINGS contexts with no RNS connection will not be updated, and so those agents will not be reachable by executing *callAgentMethod(String targetAgent, String methodName, Object[] args)*, and therefore a *CommunicationException* will be thrown if an agent tries to communicate with any of those agents executing in GeoSPRINGS contexts with no connection to an RNS; only when the mobile agent is executing in a GeoSPRINGS context under the administrative domain of an RNS, the corresponding dynamic proxies will be updated (as long as those dynamic proxies are also within contexts with RNS connection). These limitations could be overcome by supporting the use of an RNS also in ad hoc mode. However, this is challenging and could be addressed as future work.

3   Preliminary Experimental Evaluation

To test GeoSPRINGS, we have developed an Android application using the Android SDK. The application offers the following functionalities:

– *Create a local GeoSPRINGS context*. Creating a local GeoSPRINGS context on the mobile device is a first step needed to use the rest of the functionalities. As shown in Fig. 1(a), the user can specify a name for the GeoSPRINGS context, a port number to use for communications, and the address of the RNS where the GeoSPRINGS context will be linked to. In case the RNS specified is not reachable, due to the absence of Internet connection, the application will show a warning and automatically retry the connection later.

- *Observe information related to the local GeoSPRINGS context.* Through the "Context" tab, shown in Fig. 1(b), we can see logs containing information about the local GeoSPRINGS context and the agents executing there.
- *Activate and deactivate the WiFi Direct discovery service.* As the use of WiFi Direct consumes energy, we offer the possibility to disable the discovery service of nearby devices. This can be done through the "WiFi Direct" tab, shown in Fig. 1(c), which also shows the devices located nearby.
- *Geolocalize the GeoSPRINGS contexts in the current region and the nearby GeoSPRINGS contexts.* Through the "Map" tab, shown in Fig. 2, we can visualize on a map the GeoSPRINGS contexts known by the mobile device: the current context is in red, those connected to the RNS in blue, and nearby ones detected by using WiFi Direct in green. Google's API is used to update on the map the location of the local GeoSPRINGS context periodically.

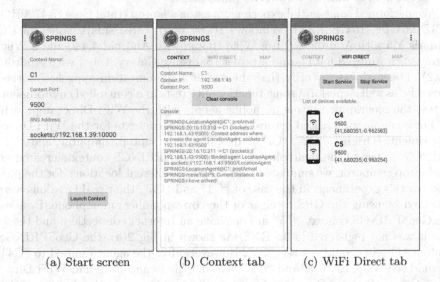

(a) Start screen   (b) Context tab   (c) WiFi Direct tab

Fig. 1. Android application developed for field testing

The most important difficulties found in the implementation are related to the management of WiFi Direct using the Android framework [2]. This requires obtaining an instance of the *WifiP2pManager* class belonging to the Android library *android.net.wifi.p2p*. This class provides methods to discover peer devices (*discoverPeers*, to scan for available WiFi peers, and *discoverServices*, to look for peers that support certain services) and establish connections with other devices. Besides, it is necessary to create a *BroadcastReceiver* in order to receive event notifications and react to them; for that purpose, an object of the class *IntentFilter* is created and the events or intents that need to be detected (e.g., WIFI_P2P_CONNECTION_CHANGED_ACTION,

WIFI_P2P_PEERS_CHANGED_ACTION) are added to that IntentFilter. Besides, it is also necessary to announce key service information of the local GeoSPRINGS context (context name, port, latitude and longitude of the mobile device) to other nearby devices through the *WifiP2pManager*; specifically, the method *addLocalService* is used to add an instance of *WifiP2pServiceInfo* (class that stores service information to advertise in a WiFi peer-to-peer network). WiFi Direct is based on asynchronous callbacks and the discovery service executes in the background. So, it is necessary to notify the main program when new information is available and therefore the Graphical User Interface (GUI) must be updated: the *sendBroadcastMessage* method is called and the notification is received by a *BroadcastReceiver* in the main program, that will update the GUI accordingly. We have noticed that the service discovery sometimes fails when a service (instance of *WifiP2pDnsSdServiceInfo*) has been published before the discovery starts; so, we create a *TimerTask* to publish the services periodically, with a certain update period, which may lead to some delays when establishing WiFi Direct connections, due to the need to wait for the service publication.

We performed some initial experiments using a laptop (Intel Core i7-4720HQ, CPU 2.6 GHz, 16 GB RAM, Windows 10 64 bits), three smartphones (a BQ Aquaris M5 with Android 6.0.1, a Wiko Bloom with Android 4.4.2, and a Wiko Sunny with Android 6.0.1), and a tablet (Samsung Galaxy Tab 4 with Android 5.0.2). The goal was to verify that the socket communication approach worked correctly, as well as perform some functionality tests in a controlled environment, to test the geographic routing of mobile agents with the WiFi Direct approach.

For example, we created several GeoSPRINGS contexts for the WiFi Direct experiment: "Origin", "C1", "C2" and "C3" on the laptop computer, and "C4" and "C5" in two of the smartphones. For the GeoSPRINGS contexts created on the laptop computer, we simulated that they had different locations, for the purposes of this experiment; in the case of "C4" and "C5", their real locations were obtained by using the GPS receiver of the corresponding smartphone. Besides, the GeoSPRINGS context "C5" did not have an Internet connection, and therefore it was not registered in an RNS. As shown in Fig. 2(a), the GeoSPRINGS context "Origin" could not see "C5". However, after the agent jumped to "C4", it could see "C5", as "C4" and "C5" were neighbors and both had WiFi Direct activated, as shown in Fig. 2(b). After establishing the connection between "C4" and "C5", the mobile agent was able to move to "C5" successfully.

We also performed a first scalability test using only the laptop computer. For this experiment, we varied the number of agents created between 100 and 500. Each agent selected randomly another agent as its peer and it repeated continuously a cycle where it called its peer and then moved randomly to another GeoSPRINGS context, until it had performed 20 iterations, and then it stopped moving and calling its peer. The total execution time for the test varied between about 20 s (with 100 agents) and 35 s (with 500 agents). The invocation time varied from about 110 ms to about 350 ms. Overall, we observed a quite linear behavior. Nevertheless, this experiment is still preliminary. Additional experiments in other scenarios, and using mobile devices with different connectivity options, are required.

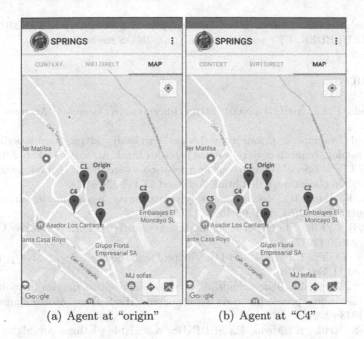

(a) Agent at "origin"   (b) Agent at "C4"

Fig. 2. Scenario for the test with WiFi direct (Color figure online)

4   Conclusions and Future Work

In this work-in-progress paper, we have presented GeoSPRINGS, a mobile agent platform with geographic capabilities. The platform can be executed on Android devices as well as on desktop computers and laptops with a JVM. It offers communication and migration methods that are location-aware, allowing a mobile agent to move to a geographic area or to call a method on all the agents located within a certain geographic area. Communications are based on the use of sockets and a proxy pattern, and so the platform does not depend on specific RMI implementations available for Android devices, which are usually subject to some limitations. Finally, we would like to emphasize that GeoSPRINGS can adapt the offered functionality depending on the connection available on the mobile device, supporting direct ad hoc communications through WiFi Direct.

The preliminary experiments performed serve as a proof of concept and show promising results. However, a more in-depth experimentation in a variety of scenarios would be required to thoroughly test the platform and its limitations from a practical point of view; for example, in some experiments WiFi Direct behaved unreliably, so it is relevant to study this issue in more detail. Besides, the platform could be extended with more advanced hopping strategies to perform a more intelligent geographic routing when extra information (e.g., expected trajectories of mobile devices) is available. Finally, more geographic-related methods could be added to support not only GPS locations but also semantic locations.

Acknowledgments. This work has been supported by the project TIN2016-78011-C4-3-R (AEI/FEDER, UE) and DGA-FSE (COS2MOS research group).

References

1. Andrade, F.S.: LipeRMI (2006). http://lipermi.sourceforge.net. Accessed 11 Feb 2018
2. Android Developers: Connecting devices wirelessly. https://developer.android.com/training/connect-devices-wirelessly/index.html. Accessed 11 Feb 2018
3. Android Developers: Making your app location-aware. https://developer.android.com/training/location/index.html. Accessed 11 Feb 2018
4. Bagga, P., Hans, R.: Mobile agents system security: a systematic survey. ACM Comput. Surv. **50**(5), 65:1–65:45 (2017)
5. Balan, R.K., Flinn, J.: Cyber foraging: fifteen years later. IEEE Pervasive Comput. **16**(3), 24–30 (2017)
6. Google Code: RipeRMI (2011). https://code.google.com/archive/p/ripermi. Accessed 11 Feb 2018
7. Ilarri, S., Mena, E., Illarramendi, A.: Location-dependent queries in mobile contexts: distributed processing using mobile agents. IEEE Trans. Mob. Comput. **5**(8), 1029–1043 (2006)
8. Ilarri, S., Trillo, R., Mena, E.: SPRINGS: a scalable platform for highly mobile agents in distributed computing environments. In: Fourth International WoWMoM Workshop on Mobile Distributed Computing (MDC), 5 pp. IEEE (2006)
9. Lange, D.B., Oshima, M.: Seven good reasons for mobile agents. Commun. ACM **42**(3), 88–89 (1999)
10. Milojicic, D., Douglis, F., Wheeler, R.: Mobility: Processes, Computers, and Agents. ACM, New York (1999)
11. Mouradian, C., Naboulsi, D., Yangui, S., Glitho, R.H., Morrow, M.J., Polakos, P.A.: A comprehensive survey on fog computing: state-of-the-art and research challenges. IEEE Commun. Surv. Tutorials **20**(1), 416–464 (2017).
12. Satyanarayanan, M., Bahl, P., Cáceres, R., Davies, N.: The case for VM-based cloudlets in mobile computing. IEEE Pervasive Comput. **8**(4), 14–23 (2009)
13. Spyrou, C., Samaras, G., Pitoura, E., Evripidou, P.: Mobile agents for wireless computing: the convergence of wireless computational models with mobile agent technologies. Mob. Netw. Appl. **9**(5), 517–528 (2004)
14. Taleb, T., Ksentini, A.: Follow me cloud: interworking federated clouds and distributed mobile networks. IEEE Netw. **27**(5), 12–19 (2013)
15. Trillo, R., Ilarri, S., Mena, E.: Comparison and performance evaluation of mobile agent platforms. In: Third International Conference on Autonomic and Autonomous Systems (ICAS), pp. 41–46. IEEE (2007)
16. Urra, O., Ilarri, S., Trillo-Lado, R.: An approach driven by mobile agents for data management in vehicular networks. Inf. Sci. **381**, 55–77 (2017)
17. Wi-Fi Alliance: Wi-Fi Direct. https://www.wi-fi.org/discover-wi-fi/wi-fi-direct. Accessed 11 Feb 2018
18. Yu, W., Liang, F., He, X., Hatcher, W.G., Lu, C., Lin, J., Yang, X.: A survey on the edge computing for the Internet of Things. IEEE Access **6**, 6900–6919 (2017).
19. Yus, R., Mena, E., Ilarri, S., Illarramendi, A.: SHERLOCK: semantic management of location-based services in wireless environments. Pervasive Mob. Comput. **15**, 87–99 (2014)

Extraction of Usage Patterns for Land-Use Types by Pedestrian Trajectory Analysis

Mehdi Jalili[1], Farshad Hakimpour[1]([✉]) [ID],
and Stefan Christiaan Van der Spek[2]

[1] School of Surveying and Geospatial Engineering,
University of Tehran, Tehran, Iran
{mehdi.jalili94, fhakimpour}@ut.ac.ir
[2] Department of Urbanism, Delft University of Technology,
Delft, The Netherlands
S.C.vanderSpek@tudelft.nl

Abstract. Research on moving objects and analysis of movement patterns in urban networks can help us evaluate urban land-use types. With the help of technologies such as global positioning systems, spatial information systems and spatial data the study of movement patterns is possible. By understanding and quantifying the patterns of pedestrian trajectories, we can find effects of and relations between urban land-use types and movements of pedestrians. Understanding urban land-use and their relationships with human activities has great implications for smart and sustainable urban development. In this study, we use the data of various urban land-use types and the trajectory of pedestrians in an urban environment. This paper presents a new approach for identifying busy urban land-use by semantic spatial trajectory in which urban land-uses are assessed according to the pedestrian trajectories. Undoubtedly, the extraction of popular urban land-uses and analysis of the association between popular places and the spatial and semantic movement allow us to improve the urban structure and city marketing system. In this regard, for semantic analysis of urban land-use, all stop points are extracted by a time threshold and they are enriched according to semantic information such as age, occupation, and gender. We examine if and how habits of using land-use types depend on qualities such as age, gender and occupation. For analysis of effects of various urban land-use types, all stop points near each urban land-use are detected. Determining what type of urban land-use cause pedestrian traffic and high absorption coefficient and what relation such high traffic has with semantic information such as age, occupation and gender. By clustering the stop points, the results indicate that stop at urban networks for each gender have a spatial correlation. Also, the results show that some urban land-use types have high traffic and we have a correlation with some semantic information such as age, gender and occupation.

Keywords: Trajectory analyze · Semantic spatial information
Clustering and urban land-use

© Springer International Publishing AG, part of Springer Nature 2018
M. R. Luaces and F. Karimipour (Eds.): W2GIS 2018, LNCS 10819, pp. 61–76, 2018.
https://doi.org/10.1007/978-3-319-90053-7_7

1 Introduction

Movement of objects in which time and location changes are recorded as a string of points as paths, has been subject of attention by researchers and is called trajectory. Trajectories include the sequence of spatial coordinates and time. Trajectory analysis allows us to extract patterns of movement and use this knowledge to manage our environment [1]. This knowledge can be used in a variety of studies, such as location-based services [2], crisis management [3], traffic management [4], and urban planning [5]. Researchers in different sciences use trajectory data to predict the route, for example, meteorologists use to find out the storm path or the direction of the movement of clouds and computer specialists use trajectory to find out the route of computer viruses [6]. Traffic and urban engineers use vehicle traffic information to determine the location of traffic monitoring and traffic forecasting systems. In recent years, many technologies have been developed to collect movement data, which can be used in spatial and temporal analysis [7].

With the advent of data collection technologies such as the global positioning system (GPS), space and aerial imagery and mobile mapping, it is possible to collect large amounts of spatiotemporal data and the tendency to analyze the moving objects is further increased. Considering high volume of data from which useful information must be extracted, the need for methods to extract useful information is well understood. In urban environments, using patterns of movements, we can extract knowledge such as high-crowded places, paths of commuters, time-spent in each location, frequently visited places. By acquiring such knowledge, we can then improve urban environment through minor changes to the urban structure, such as the construction of a fly over or a new route [8].

Technological advancements in trajectory data collection are far more than current methods for extracting patterns in trajectory data. Although much progress has been made in extracting patterns of motion, ignoring semantics is an important issue in trajectory analysis. Purves et al. [9] have presented a list of techniques for enrichment of trajectory data by collecting semantics:

1. along with the collection of trajectories;
2. by trajectories with descriptions of each location;
3. using physical characteristics of individuals.

We have used cases 1 and 2 to enrich our trajectories. The fact that different land-use in the city and the social characteristics of individuals have an effect on the choice of type of transportation in daily traffic [10]. Analysis the impact of pedestrian movement in the city and the impact of urban land-use on people movement is an important issue in today's research, which was solved using operational data and questionnaires in the past [11, 12]. In today's studies, trajectories use to identify these patterns from raw data and enrich these data with semantic information for the production of semantic trajectories [13]. Semantic trajectories are used in various topics such as navigation [14], understanding the social structure of the trajectories [15], etc.

City center have primary role in surveying local residents and attracting people with shopping, leisure and etc. an attractive land-use types may correlate with people's occupation, gender and age. Integrating trajectory with semantic information such as a age, gender and occupation allows us for calculating relation between land-use types and those semantics. We are interested in how gender, occupation and age might affect movement patterns. We evaluate our framework on a data set of GPS trajectories from 325 volunteers in Delft towns who were continuously tracked in four daily movements for a week. We enrich these trajectories with contextual information like a age, gender and occupation.

The movement patterns of people are connected to how people affect semantic information of places and how semantic information affects places [16]. As Kwan argued in his article [17], spatial experiences of an individuals are not only affected by the places where they live, but also depends on the visited places and the length of stop times. In this research, we introduce a different approach in which the purpose is studying various urban land-use types in relation to pedestrian trajectories. To that end, land-use types with high traffic rate according to the semantic trajectories are extracted. In this research, we analyze the correlation of stop points and the urban land-use types in four consecutive days. The main question is what kind of people visit what types of land-uses.

2 Related Work

As mobile phones are always at hand, mobile phone data are detected as good proxies of people's activities [18]. Understanding citywide human activities and the urban dynamics behind them face big challenges. Torrens et al. worked in the identification of pedestrian behaviors with machine learning [19]. Hu et al. classified different motion types with neural networks (Self-Organizing Maps) [20].

Significant places can be identified from GPS trajectories using machine or learning clustering [21, 22]. Van der Hoeven et al. present a high correlation between duration of stay in city and amount of money spend in this stay [23]. Van der Spek et al. motivated on public space usage, experiences and quality within urban cores. They tried to compile empirical evidence obtained from two GPS tracking studies demonstrating the public space use patterns of three different user groups. They present that streets with good spatial quality attract a variety of people who may stay for a longer time on the streets or visit destinations along the streets [24].

Although many approaches have been developed for trajectory semantic enrichment, i.e. Bayes activity implications [25] and transportation segmentation [26], enriching activity information using mobile phone data is not easy. Recently, combining mobile phone data with urban land-use data, [27].

Umair et al., In their article [28], used the trajectories to find important places using the compression analysis method to identify the points near the important places, in which semantic information has been ignored developed a probability approach for extracting diurnal activities (including in-home activities, working, shopping, café, restaurant and leisure) from mobile phone data and analyzed citywide activity patterns. Fusing mobile phone data with data from other sources is another promising approach. Using time series phone call records, [29] presented a functional network approach to automatically calculate four types of land use (residential, business, logistics/industrial and nightlife). Recently, Ahas et al. [30] detected spatio-temporal differences in diurnal activities in cities. Following the aggregated rhythm, social time (the time use difference) from human activities was defined instead of the standard solar time. Their results suggest global temporal dynamics in people's daily lives that cannot be ignored.

3   Modeling Theoretical and Research Process

In this section, we explain the theoretical foundations needed in this research.

3.1   Pedestrian Trajectories

Today's urban transport and traffic is not just considering vehicle movement, but the goal is to look at non-motorized transportation and improve the system of pedestrian network. The relationship between land-use types and semantics of pedestrian trajectories can help us understanding urban environments and high traffic and popular land-use types. In Eq. 1, a series of trajectories is introduced; in which p is the series of points of a trajectory and t is the time series of a trajectory.

$$\tau_{Gps} = \langle (p,t)_1, \ldots, (p,t)_n \rangle \tag{1}$$

3.2   Stop Points Representation

In the analysis of raw trajectories, the stop points have semantic information unlike other points in the trajectories. In a trajectory, the locations where the users have stopped is considered as a point of interest (POI). Stopping points occur on the trajectory due to the presence of favorable locations in the passageways. Presence of these points in the passageway increases the attractiveness of the route for people. Each of the stop points has spatial coordinates $p.s_x$ and $p.s_y$ and beginning time $t.b$ and end time of stop points $t.e$.

Stop points carry more semantics beyond raw GPS points, and allow us to filter the places where a user only passed by, e.g., crossroads. In this work, stop point detection extracts important places from massive mobile phone positioning data. Mobile phone records of an individual are first sorted by time and connected as a spatiotemporal trajectory. Then, if two consecutive records are at the same location, in other words, the person does not move, a Stop points is found, such as p4 in Fig. 1.

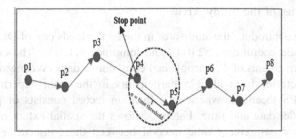

Fig. 1. A GPS trajectory and a stay point

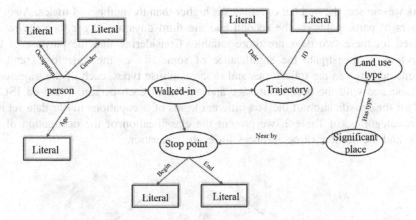

Fig. 2. Trajectory enriching and relation between stop points and urban land-use.

Semantic trajectories

The analysis of raw trajectories that are limited to spatial and temporal data faces shortcomings. The stop points in contrast to the raw points in the trajectories have semantic information for each person. Many researchers have mentioned the lack of information in raw trajectories for the applied interpretation of analysis [31, 32]. Therefore, the process of extracting high-level information from a low-level data set (time-space) requires enriching these data with contextual data [33]. Today, the semantic presentation of trajectories is always considered in research [13]. In the semantic trajectories, the stop points of the trajectories are labeled with semantic information. In this process, each trajectory is labeled with contextual information, as shown in Fig. 2.

4   Study Area

To implement our proposed approach, we use trajectories and urban land-use types as data sources describing the environment.

4.1   Trajectories of the Study Area

In this section, we introduce the study area in the Netherlands city of Delft. This area is located on latitude coordinates 52.0119 and longitude 4.43026. The collected trajectories contain movements of 325 people in 4 consecutive days (Wednesday to Saturday in 2009). Trajectories from all 325 participants in the main experiment comprise 465,610 raw GPS locations, where each location record consists of participant ID, latitude, longitude, date and time. Figure 3 shows the spatial extent of the collected trajectories. The measurement time interval between the consecutive points of the trajectories is 5 s.

Average age of the participants taking part in trajectory data collection as provided in Table 1.

As we can see, the number of females is higher than the number of males. Also, the numbers of participants on the second and the third days are larger then the results obtained for these two days are more reliable. Considering that the purpose of this research is to investigate the significance of semantics of the trajectories and the relationship between the trajectories and urban land-use types, each of the trajectories are associated with the occupation, age and gender of each person. We used **ISCO**[1] standard for classification of the 164 different kinds of occupations in our data set into eleven categories. In Table 2, we present the classification of the occupation of the people and the number of occurrences for each occupation.

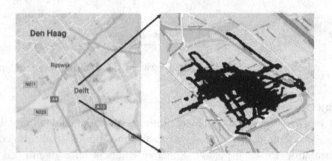

Fig. 3. Trajectories collected in the GPS travel survey

Table 1. Participants in the GPS survey.

Day	Male count	Avg. age	Female count	Avg. age	Total
One	18	48.3	49	43.2	67
Two	44	54.8	59	47.2	103
Three	43	46.2	65	44.7	108
Four	23	47.4	24	52	47
Total	128	49.66	197	45.96	325

---

[1] http://www.ilo.org/public/english/bureau/stat/isco/.

Table 2. Participant occupation in study area.

Occupation	Count
Pension	53
Science professionals	49
Housewife	36
Health associate	36
Official	33
Manager	31
Legal	23
Machine operators	18
Elementary occupations	17
Sale works	15
Student	14
Total	325

4.2 Urban Land-Use of the Study Area

Additional datasets are needed to model the potential interest of pedestrians in a land-use. Volunteering data has been used for various urban applications and geospatial research for example Open Street Map data is used as a component of an urban space network [34]. In Fig. 4 the spatial distribution of urban land-uses of the study area is presented.

Urban land-use types of study area have different spatial distributions. Table 3 provides information of eleven land-use types we considered in this research and frequency of each land-use type.

Fig. 4. OSM urban land-use in study area.

Table 3. Land-use type frequency in study area.

Land-use type	Frequency
Shop	41.5
Restaurant	19.5
Café	14.5
Office	5.4
Healthcare service	5.02
Art (gallery or museum)	4.4
Public transport	4.2
Hotel	2.12
Educational	1.35
Gym	0.96
Holy place	0.77

5 Methodology

In this section, we explain the implementation and analysis of the spatial semantic trajectories for calculation of the relation between stop points and urban land-use types. A data integration framework is presented to discover urban attractive land-use by extracting stop points in trajectories and enriching land-uses. An attractive land-use is first detected by time spent in trajectories nearby. The presented approach infers urban land-use and their diurnal dynamics by combining massive pedestrian trajectory data and land-use types. In this research a database is used to store and manage trajectory data. The solution framework includes three phases described in the reminder of this section.

5.1 Extracting Stop Points

In analyzing trajectories, we consider a trajectory as a sequence of points. This section aims to find stop points in each of trajectory. For extracting the stop points, we use a time window that examines each point of the trajectories. In other words, to determine the start and end points of stop points, the algorithm looks for time gap between each point and the next point. It scans each trajectory using moving windows. In this analysis, $\sigma = 90$ s threshold is used. Time difference between each point in trajectories is compared to the next point and if it is more than the specified time window, it is opted as the stop point. After extracting the stop points the time difference between the stop point and the next point is considered as the time gap. In total, 1795 stop points were detected in our dataset. On average, one person holds 6.79 stop points in a day. In Fig. 5 the stop points are shown in four consecutive days.

This case study aims to provide an initial insight on how the investigation of movement patterns of demographic groups can provide informative data for urban planners. On the other hand, the magnitude of downtown area to midtown area ratios also varies substantially for different cities. A higher downtown to midtown area ratio indicates a more concentrated planning pattern–such as Delft. This may be the result of

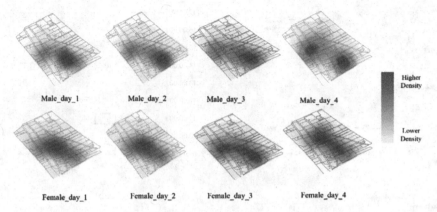

Fig. 5. Kernel density distribution of gender groups in city Delft.

various factors, such as the spatial distribution of work opportunities, residential areas, education institutes, transportation infrastructure, recreational facilities etc. To validate these findings, we also plotted the kernel density estimation of the stop points in Fig. 5 for male and female in the interval time of 9 am to 6 pm (search radius = 344 m). Although female groups indicate clustering in the city, whereas the male groups show more spread patterns in the southern areas.

As you can see, the stop points have a certain distribution pattern for both genders. The cluster patterns for the same gender show similarity in different days. On weekdays Men's tend to stop at the southern part of the study area, and women tend to stop in the central part of the study area. On the fourth day, this pattern is slightly different for men, which may be due to the weekend. Female pedestrians have more stop and appear to visit land-uses more frequently than males. However, females have more access to the city center areas possibly due to home duties such as shopping.

5.2 Time-Base Trajectory Clustering

One-dimensional K-means clustering is used to recognize the temporal patterns of pedestrian movements. By clustering of times for stop points in each day, we can detect busy times.

As previously mentioned, the purpose of this study is to analyze the urban land-use types and extract busy times also finding out the attributes of people visiting urban land-uses at those times. After extracting the stop points of the trajectories for four consecutive days and the semantic enrichment of these points, the frequency of each point is given in terms of the time elapsed and the time of occurrence in Fig. 6. As seen in some hours, the stop frequency is high. Stop points and their times are aggregated for each land-use type then busy urban land-uses are inferred and the diurnal dynamics of those land-uses are revealed.

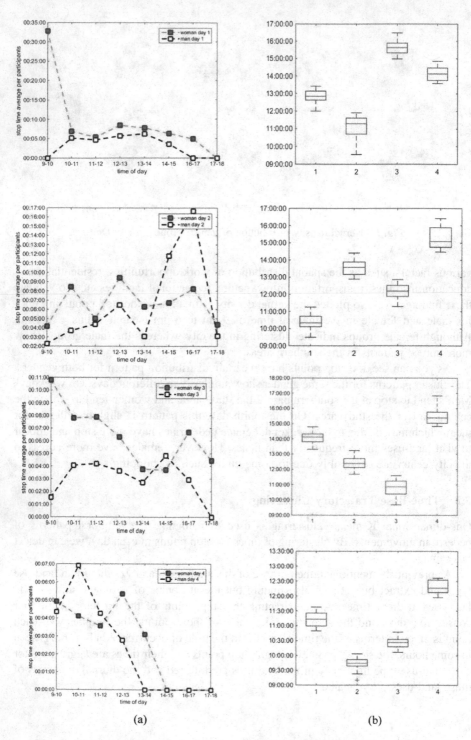

Fig. 6. Stop point frequency in four days based on gender and stop time clustering.

5.3   Analysis of Urban Land-Use Types with Semantic Stop Points

For analysis of urban land-use types to find out what kinds of stop points cause high traffic, the tables of semantic stop points and urban land-use are entered into a relational database (the relations are shown in Fig. 7).

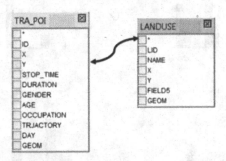

Fig. 7. Stop point and urban land-use in relational database.

For analysis of urban land-use, the tables for stop points and urban land-use should be joined. In this analysis, all points within 10 m of each land-uses type are required. After joining two tables in the database, a spatial threshold of 10 m is used to extract how many stop points are nearby each land-use. First, we calculated the proximity table for semantic stop point and land-uses types. As mentioned in Sect. 1, human activities are restricted by a multitude of factors, including high-class-individual factors such as the built environment and individual-level factors such as occupation, age and gender. Based on these factors, it is feasible to identify particular patterns for population groups categorized by social attributes. This section focuses on analyzing how individual and individual factors affect the distribution of mobility patterns.

To identify significant land-use types, we calculate the frequency of reoccurrence and the amount of time spent in an urban land-use for all the semantic stop points. The most frequently visited place with the longest total duration of visits is considered to be the place with high absorption coefficient.

For each type of land-uses the count of semantic stop points is calculated based on day and gender. In other words, land-use traffic is firstly classified according to the day and each of these counts for land-uses are divided by gender. We calculated the average time per day spent at popular land-use types per participant. Most participants spent a reasonable amount of time at popular land-uses (mostly between 5 to 30 min) and most people, as expected, stayed at café and restaurant longer during the weekdays. We also investigated the average times per day spent in shop, café and restaurant, broken down by days and gender (Fig. 8). Interestingly, the participants spent a relatively large amount of time in their café and restaurant, regardless of the day1, day2, day3 and day4.

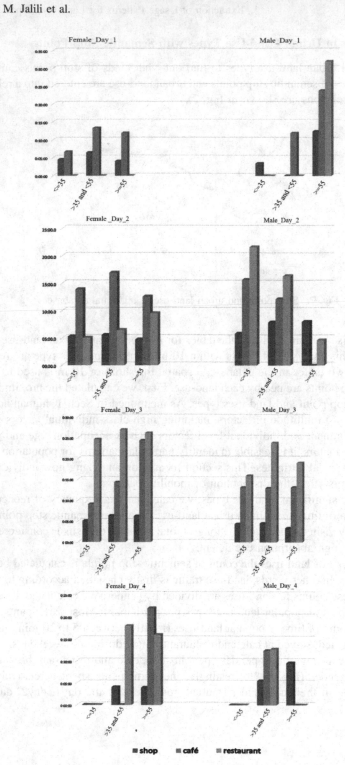

Fig. 8. Busy urban land-use types based on age and average of stop point.

6 Assessment

In this section, first spatial correlation of the stop points and urban land-use types are presented to expose the pattern of stop points for pedestrians in each of the urban land-uses. The purpose of this section is to calculate the correlation between urban land-use and semantic stop points in four days. In calculation of the correlation we used existing software tools that do not take into account spatial autocorrelation. Moreover, we present the percentage of usage for each land-use type according to the occupation.

6.1 Correlation Between People Distribution and Land-Use Type

In this section we want to expose correlation between spatial distribution of stop points (as shown in Fig. 5) and spatio-semantic of land-use types. Table 4 provides correlation between parameters. As one can see, there is a higher correlation between male and female stop points and shop, café, public transport and restaurant compared to art place, healthcare service. Yet, art galleries and museums have higher correlation in the weekend for both genders. There is also a slightly higher correlation between female stop points and shops in average compared to male stop points.

Table 4. Correlation between gender and land-use types.

		Shop	Café	Public transport	Restaurant	Health care	Art	Office
Male	Day1	0.621	0.623	0.621	0.429	0.382	0.063	0.397
	Day2	0.887	0.887	0.887	0.792	0.497	0.218	0.631
	Day3	0.783	0.783	0.783	0.583	0.399	0.153	0.551
	Day4	0.749	0.749	0.749	0.653	0.576	0.204	0.239
Female	Day1	0.719	0.720	0.720	0.508	0.325	0.086	0.460
	Day2	0.777	0.757	0.687	0.673	0.582	0.318	0.557
	Day3	0.907	0.907	0.907	0.794	0.401	0.432	0.638
	Day4	0.843	0.843	0.843	0.668	0.456	0.528	0.472

6.2 Semantic Correlation Between Urban Use and Semantic Stopping Points

In this section, the correlation between occupation type and land-use types is calculated in 4 consecutive days. Table 5 shows result of correlation. Most people with different occupations have used the shop. As we can see, some of the urban land-use types attract more attention independent of pedestrians' type of occupation–e.g. shop. In Contrast office land-use type is used significantly more by managers and then officials and science professionals.

Table 5. Percentage of usage of each land-use type according to the occupation.

	Shop	Café	Public transport	Restaurant	Health care	Art	Office	Total
Pension	57.07	15.76	5.43	13.04	4.35	2.72	1.63	100
Science professional	69.41	6.47	2.94	6.47	1.18	1.18	12.35	100
Housewife	51.81	10.36	4.15	15.54	3.63	6.22	8.29	100
Health associate	60.49	9.26	6.17	14.20	4.32	3.70	1.85	100
Official	55.08	7.14	0.00	17.46	6.35	1.59	12.38	100
Manager	60.18	1.80	3.60	6.31	2.70	0.00	25.41	100
Legal	73.44	7.81	3.13	12.50	0.00	0.00	3.13	100
Mechanic	54.35	8.70	10.87	19.57	4.35	0.00	2.17	100
Sale work	66.67	10.00	0.00	20.00	3.33	0.00	0.00	100
Student	58.82	23.53	0.00	17.65	0.00	0.00	0.00	100

7 Conclusion

Our objective in this research is to develop a methodology for measuring stop points patterns of urban pedestrian, which could be useful in urban monitoring tools and city marketing solutions. In the era of big data, massive human tracking data are available (e.g. mobile phone data, GPS data, social media data etc.) but lack of semantics is a big issue in related of studies. Understanding urban land-use and pedestrian movement is essential for the planning, managing and governing of a city. Understanding how habit of using urban land-use types vary across space and time is very challenging. In order to improve city center environment and obtain correlation between semantic trajectory and land-use types, knowing how different social group tend to use urban land-use types are fundamental. This article presents an approach to uncover urban popular land-use types and their diurnal dynamics by using pedestrian trajectory and urban land-use. First the stop points are detected using temporal moving window, then all of these points enriched with semantics such as age, gender, occupation and date. We classified all of the occupations based on ISCO standard. We used OpenStreetMap land-use types. Then we classified stop point based on kernel density estimation and it shows the stop points have a certain distribution pattern for both genders. The cluster patterns for the same gender show similarity in different days. The stop time of human activities in each urban cell is calculated. The results also reveal that urban land-use have different pattern based on gender and age of people hour by hour. Although this research is at its early stages, it shows that land-use types have correlation with kind of occupations. Also, this study reveals busy time intervals when people stop more and the type of urban land-use that have correlation with semantic trajectory.

References

1. Classifying pedestrian movement behaviour from GPS trajectories using visualization and clustering: Annals of GIS, vol. 20(2). http://www.tandfonline.com/doi/full/10.1080/19475683.2014.904560. Accessed 28 Apr 2017
2. Millonig, A., Gartner, G.: Identifying motion and interest patterns of shoppers for developing personalised wayfinding tools. J. Locat. Based Serv. 5(1), 3–21 (2011)
3. Zheng, X., Zhong, T., Liu, M.: Modeling crowd evacuation of a building based on seven methodological approaches. Build. Environ. 44(3), 437–445 (2009)
4. Castro, P.S., Zhang, D., Li, S.: Urban traffic modelling and prediction using large scale taxi GPS traces. In: Pervasive Computing, pp. 57–72 (2012)
5. van der Spek, S.: Tracking tourists in historic city centres. In: Gretzel, U., Law, R., Fuchs, M. (eds.) Information and Communication Technologies in Tourism, pp. 185–196. Springer, Vienna (2010). https://doi.org/10.1007/978-3-211-99407-8_16
6. Eubank, S., et al.: Modelling disease outbreaks in realistic urban social networks. Nature 429 (6988), 180–184 (2004)
7. Kwan, M.-P., Neutens, T.: Space-time research in GIScience. Int. J. Geogr. Inf. Sci. 28(5), 851–854 (2014)
8. Van Schaick, J.: Timespace matters - Exploring the gap between knowing about activity patterns of people and knowing how to design and plan urban areas and regions (2011)
9. Purves, R.S., Laube, P., Buchin, M., Speckmann, B.: Moving beyond the point: An agenda for research in movement analysis with real data, vol. 47 (2014)
10. Ewing, R., Cervero, R.: Travel and the built environment. J. Am. Plann. Assoc. 76(3), 265–294 (2010)
11. Van Vugt, M., Van Lange, P.A.M., Meertens, R.M.: Commuting by car or public transportation? A social dilemma analysis of travel mode judgements. Eur. J. Soc. Psychol. 26(3), 373–395 (1996)
12. Wener, R.E., Evans, G.W.: A morning stroll: levels of physical activity in car and mass transit commuting. Environ. Behav. 39(1), 62–74 (2007)
13. Parent, C.: Semantic trajectories modeling and analysis. ACM Comput. Surv. 45, 1–32 (2013)
14. Sester, M., Feuerhake, U., Kuntzsch, C., Zhang, L.: Revealing underlying structure and behaviour from movement data. Künstl. Intell. 26(3), 223–231 (2012)
15. Xiao, X., Zheng, Y., Luo, Q., Xie, X.: Inferring social ties between users with human location history. J. Ambient Intell. Humaniz. Comput. (2012)
16. Palmer, J.R.B., Espenshade, T.J., Bartumeus, F., Chung, C.Y., Ozgencil, N.E., Li, K.: New approaches to human mobility: using mobile phones for demographic research. Demography 50(3), 1105–1128 (2013)
17. Kwan, M.-P.: Beyond space (As We Knew It): toward temporally integrated geographies of segregation, health, and accessibility. Ann. Assoc. Am. Geogr. 103(5), 1078–1086 (2013)
18. Tranos, E., Nijkamp, P.: Mobile phone usage in complex urban systems: a space–time, aggregated human activity study. J. Geogr. Syst. 17(2), 157–185 (2015)
19. Torrens, P., Li, X., Griffin, W.A.: Building agent-based walking models by machine-learning on diverse databases of space-time trajectory samples. Trans. GIS 15, 67–94 (2011)
20. Hu, W., Xie, D., Tan, T.: A hierarchical self-organizing approach for learning the patterns of motion trajectories. IEEE Trans. Neural Netw. 15(1), 135–144 (2004)
21. Liao, L., Fox, D., Kautz, H.: Extracting places and activities from GPS traces using hierarchical conditional random fields. Int. J. Robot. Res. 26(1), 119–134 (2007)

22. Rodrigues, A., Damásio, C., Cunha, J.E.: Using GPS logs to identify agronomical activities. In: Huerta, J., Schade, S., Granell, C. (eds.) Connecting a Digital Europe Through Location and Place. LNGC, pp. 105–121. Springer, Cham (2014). https://doi.org/10.1007/978-3-319-03611-3_7

23. Van der Hoeven, F.D., Van der Spek, S.C., Smit, M.G.J.: Street-Level Desires, Discovering the City on Foot: Pedestrian Mobility and the Regeneration of the European City Centre (2008)

24. Van der Spek, S.C., Van Langelaar, C.M., Kickert, C.C.: Evidence-based design: satellite positioning studies of city centre user groups. Proc. Inst. Civ. Eng. Urban Des. Plan. 166(4), 206–216 (2013)

25. Gong, L., Liu, X., Wu, L., Liu, Y.: Inferring trip purposes and uncovering travel patterns from taxi trajectory data. Cartogr. Geogr. Inf. Sci. 43(2), 103–114 (2016)

26. Shin, D., et al.: Urban sensing: using smartphones for transportation mode classification. Comput. Environ. Urban Syst. 53(Supplement C), 76–86 (2015)

27. Widhalm, P., Yang, Y., Ulm, M., Athavale, S., González, M.C.: Discovering urban activity patterns in cell phone data. Transportation 42(4), 597–623 (2015)

28. Umair, M., Kim, W.S., Choi, B.C., Jung, S.Y.: Discovering personal places from location traces. In: 16th International Conference on Advanced Communication Technology, pp. 709–713 (2014)

29. Lenormand, M., et al.: Comparing and modelling land use organization in cities. R. Soc. Open Sci. 2(12), 150449 (2015)

30. Ahas, R., et al.: Everyday space–time geographies: using mobile phone-based sensor data to monitor urban activity in Harbin, Paris, and Tallinn. Int. J. Geogr. Inf. Sci. 29(11), 2017–2039 (2015)

31. Guc, B., May, M., Saygin, Y., et al.: Semantic annotation of GPS trajectories. In: Proceedings of the AGILE 2008, CD-ROM, p. 9 (2008)

32. Spaccapietra, S., Parent, C., Damiani, M.L., de Macedo, J.A., Porto, F., Vangenot, C.: A conceptual view on trajectories. Data Knowl. Eng. 65(1), 126–146 (2008)

33. van Hage, W.R., Malaisé, V., de Vries, G., Schreiber, G., van Someren, M.: Combining ship trajectories and semantics with the simple event model (SEM). In: Proceedings of the 1st ACM International Workshop on Events in Multimedia, New York, pp. 73–80 (2009)

34. An Outlook for OpenStreetMap. https://www.researchgate.net/publication/290522341_An_Outlook_for_OpenStreetMap. Accessed 21 Apr 2017

Hierarchical Routing Techniques
in Wireless Sensor Networks

Amine Kardi[1(✉)], Rachid Zagrouba[2], and Mohammed Alqhtani[2]

[1] Faculty of Mathematical, Physical and Natural Sciences of Tunis,
University of Tunis El Manar, Tunis, Tunisia
aminekardi@yahoo.fr
[2] College of Computer Science and Information Technology,
Imam Abdulrahman Bin Faisal University, Dammam, Saudi Arabia

Abstract. This paper studies the hierarchical routing in Wireless Sensor
Networks (WSNs) following a new in-depth classification model comprising
nine classes, discussed briefly in this study (purposes, characteristics...), in
order to provide a deep assessment facilitating the implementation of routing
protocols for researchers and protocols designers.

Keywords: WSNs · Sensor · Routing protocols · Hierarchical

1 Introduction

Owing to the advances in Micro-Electro-Mechanical System (MEMS) and wireless
communication technologies, WSNs composed of large number of resource con-
strained sensor nodes and used in a wide range of applications (military, healthcare,
surveillance....) have received tremendous attention from scientific and industrial
community in the past few years.

Various researches interested in routing in WSNs have proved that routing task
directly affects the network resources. It acts directly on the energy consumption and
therefore on the network lifetime [1, 2].

Recent researches interested in routing in WSNs, have led to the appearance of a
variety of routing protocols that aim to overcome the severe hardware and resources
constraints of nodes since the variety of applications makes a minority of routing
protocols inefficient for sensor networks across all applications [3].

The main contribution of this paper is to present an in-depth classification model of
routing protocols in WSNs in terms of precision and clarity focusing specifically on
Hierarchical routing protocols.

The rest of the paper is organized as follows: Sect. 2, presents the routing challenges and the design issues in WSNs. Section 3 presents our proposed taxonomy of
routing protocols. In Sect. 4 we discuss the hierarchical routing model. Lastly, we
conclude the paper and highlight our future work in Sect. 5.

© Springer International Publishing AG, part of Springer Nature 2018
M. R. Luaces and F. Karimipour (Eds.): W2GIS 2018, LNCS 10819, pp. 77–84, 2018.
https://doi.org/10.1007/978-3-319-90053-7_8

2 Routing Challenges and Design Issues

Sensor nodes, composed of various units (power, processing, sensing, transceiver and some additional units) as shown in Fig. 1, made a specific network architecture distinguished from classical wireless communication models (VANETs, FANETs…) by its unique and specific characteristics such as the limited energy and computational capacities, the restricted coverage capabilities, the sensitivity of links, and the intermittent connectivity.

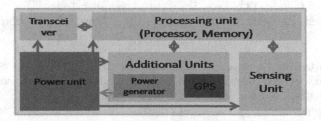

Fig. 1. Architecture of a sensor node

Hence, performing data communication and ensuring reliable multi hop communication from the target area towards the sink node in an infrastructure-free topology, while trying to enhance the lifetime of the network, is the main objective of routing protocols in WSNs [4].

Many routing protocols have been proposed for routing data in WSNs, and the presence of a global taxonomy model is necessary to analyze and classify them. We present in the next section a well detailed taxonomy, in terms of punctuality and preciseness, of routing protocols in general and especially of hierarchical protocols.

3 The Proposed Taxonomy of Routing Protocols in WSNs

Several specialized routing protocols have been proposed to conserve power and to extend the network lifetime because conventional routing protocols, designed for ad-hoc networks, are not applicable in a relevant way to WSNs due to their specific characteristics.

These protocols differ in various ways, and a global taxonomy is needed to select the most appropriate protocol for an application based on its requirements.

Based on their procedures, parameters, attributes and aspects, we proposed, a novel comprehensive and well detailed taxonomy dividing routing protocols into nine categories of routing protocols namely Application type, Delivery Mode, Initiator of communication, Network architecture, Path establishment (Route Discovery), Network topology (Structure), Protocol operation, Next Hop selection and Latency-aware and energy-efficient Routing protocols. In the rest of this section, we discuss briefly these routing paradigms.

3.1 Application Type

Routing protocols for WSNs can be classified into event-driven and time-driven protocols respectively if the data will be sent following a triggering event or periodically. These protocols can be further divided into Sink centric and Node centric protocols depending on whether decisions and sensing levels are made by sinks and sent to sensors or predefined at the end nodes. Contrary to time-driven protocols, where data are sent periodically to the sink, with a prefixed or configurable reporting period.

3.2 Delivery Mode

Temporal constraints vary from an application to another. Real Time protocols ensure successful data delivery without temporal constraints while Non Real Time-protocols are used in applications requiring a real time communication.

3.3 Network Architecture

Based on the network architecture, routing protocols can be also classified into Data centric and Position centric (Geo-centric) routing protocols. Data centric routing protocol aim to eliminate redundant messages and to refine data filling using a data naming mechanism contrary to Position centric routing protocols where nodes are position aware and data and queries are forwarded directly to a specific region to eliminate redundant data and unnecessary queries.

3.4 Initiator of Communication

In WSNs, the routing paths can be initiated by source nodes (source of sensed data) or the sink (destination of sensed data).

3.5 Path Establishment

Routing protocols in WSNs can be also classified to into three classes namely proactive, reactive or hybrid based on their routes discovering process. With proactive protocols, routing tables are generated and updated before they are needed contrary to Reactive routing protocols which generate routing table only on-demand. Hybrid model combines the two mechanisms to benefit of proactivity and reactivity.

3.6 Network Topology

Routing protocols in WSNs can be also classified into Hierarchical, Flat, Heterogeneity based, Mobility based and Geo-routing protocols according to the network structure (topology).

Hierarchical routing protocols (Cluster based) divide the network in small groups of sensor nodes called clusters with a selected Cluster Head (CH) in each group used to ensure communication in the inter and intra-cluster domain. These protocols can be also divided into three categories namely Block cluster based, Grid cluster based and Chain cluster based. This class will be studied further in the next section.

Flat routing protocols define a specific where all nodes are treated equally, have identical functionality and carry out the same tasks in gathering information.

Heterogeneity based protocols are capable to manage routing in specific WSNs topologies where various types of sensors nodes with variable characteristics are used.

Mobility based protocols are used to manage and overcome routing problems in some WSNs with frequent changes in network's structure caused by nodes/sinks mobility.

Geo-routing protocols are used when sensor nodes are aware about their positions and can be divided into position-based routing and geocasting depending on whether data is retrieved from a single source node or requires the collaboration of several sensor nodes in a particular region.

3.7   Protocol Operation

Based on their routing operations, routing protocols can be classified into: Multipath based, query based, negotiation based, QoS based and coherent based routing protocols.

Multipath routing uses multiple paths between source and destination instead of a single path to provide reliability and network robustness to node and links failures. These protocols may be further divided into four classes namely: Alternative Path Routing in which only one path is used among the maintained paths and will be changed in case of breakdown by another. Load balancing routing protocols used to divert traffic when a main link becomes over utilized to minimize the risk of traffic congestion. Energy aware multipath protocols which select the routing path that reduces energy consumption as much as and Data Transmission Reliability protocols based on simultaneous sending of multiple copies of data across multiple paths.

Query based protocols ensure the routing task by sending query packets to retrieve specific information from sensor nodes.

Negotiation based protocols which eliminate redundant data using some negotiation process between neighboring nodes based on high level data descriptors.

QoS based routing protocols take into consideration the QoS requirements in the network in addition to extending the network life time.

Routing protocols can be coherent or non-coherent depending on whether data will be locally treated before being sent to other nodes for more processing in the case of non coherent routing or only a minimum processing is done locally using coherent routing protocols.

3.8   Next Hop Selection

These protocols, based on the next-hop for the query and/or the response, can be divided into six classes namely: Broadcast based protocols in which packets are distributed to every node in the network using broadcasting mechanism. Location based protocols using geographic information to select the next. Content based protocols used in networks where the communication model is not based on contents and not on nodes addresses and next hops are inferred from data carried by the packet. Probabilistic based routing protocols where the next hop is selected randomly among available neighbors in order to increase load balancing in the network contrary to Opportunistic based protocols where the next hop is selected based on various prioritization metrics. Finally, Hierarchical based protocols, where the next hop selection follows a hierarchical-based scheme from the source to the destination.

3.9   Latency Aware and Energy-Efficient Routing

Routing protocols in WSNs can be divided into four subcategories based on the routing objectives such as latency and energy-efficiency namely: Cluster based protocols which aim to balance the efficiency on energy using a specific routing scheme based on clusters. Multipath based protocols which balance the traffic load using multiple paths as already explained. Location based protocol which use geographic information to ensure the network latency and to achieve maximum energy efficiency in the network, as explained earlier in this paper. Finally, Heuristic and swarm based protocols which are inspired by behaviors observed in nature as in ant and bee colonies aiming to achieve energy efficiency. Heuristic and swarm based protocols can be divided into four subclasses namely SB data-centric routing protocols which are Data Centric based, SB location-based protocols using location information to know separating distance between particular nodes, SB hierarchical protocols using specific structured topology inspired from nature e.g. eggs and larvae in ant colonies which are grouped into a number of small groups according to their degree of similarity and Network flow and QoS-aware protocols aiming to satisfy some QoS metrics in the network based on some algorithms inspired from nature.

4   Hierarchical Routing

Clustering is an energy efficient communication mechanism used to transmit data hierarchically from sensor to sinks and conversely. In fact, sensor nodes are separated into small groups called clusters or clumps. Each cluster has a unique cluster head (CH) elected in different ways using various strategies following energy, coverage or neighborhood criteria. Those CHs coordinate the data transmission activities of all nodes of their clumps and can directly communicate between them or with the Base Station (BS) as shown Fig. 2 [5].

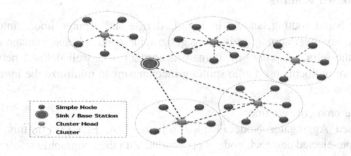

Fig. 2. Hierarchical network structure

Many hierarchical routing protocols have been proposed. Each of them setup and manage the network structure in its own way. These protocols can be classified into three categories: Block cluster based, Grid cluster based and Chain cluster based. In the following, we detail each of these categories and present its representative protocols.

4.1 Chain Based Routing

Chain cluster based routing, shown in Fig. 3, build chains of nodes that operate according to a well defined strategy where one node is selected as head to perform data transmission e.g. all the nodes in the chain can transmit and receive data from its neighbor nodes.

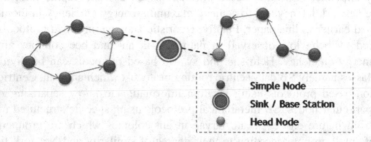

Fig. 3. Chain based routing model

Among these protocols we quote:
Track-Sector Clustering (TSC) [6]: It divides the network into concentric circular tracks and triangular sectors which transforms long chains into smaller ones to minimize unessential data transmission in order to save energy. Based on four phases namely Track setup, Sector setup and cluster-head selection, Chain construction, and Data transmission TSC proceeds as follows: it assigns each node to particular track (level) when the BS setup the sectors and selects a CH to each one. After that, TSC forms chains within each cluster by the intersection of tracks and sectors to transmit packets to the BS. Simulations show that TSC minimizes the transmission of unnecessary data but it complains of unbalanced utilization of energy throughout the entire network.

4.2 Grid Based Routing

Grid cluster based routing, shown in Fig. 4, divide and arrange nodes into equal, adjacent, and non-overlapping grids using geographic approach and manage communication in the intra and inter-grid domain according to a well defined networking strategy. These protocols use traffic splitting mechanisms to minimize the intra cluster communication cost.

Among these protocols we find:
Position based Aggregator Node Election scheme (PANEL) [7]: In this scheme, clusters are pre-elected and each node is pre-loaded with the geographic information of its cluster. As for intra-cluster communication, position based routing scheme is used

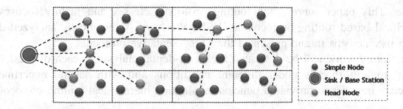

Fig. 4. Grid based routing model

for inter-cluster communication. PANEL use the concept of the reference point for CH election in each epoch (time unit) which is an energy-efficient method used to ensure load balancing and to extend the network life time but the preconfigured scheme makes it unsuitable in several cases and restricts the use of PANEL in WSNs.

4.3 Block Based Routing

Block cluster based routing, shown in Fig. 5, arrange sensor nodes in virtual block where communications and data transmissions in and between blocks follow a networking strategy.

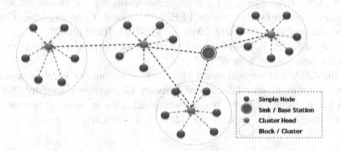

Fig. 5. Block based routing model

Among these protocols we quote:
Multi-Weight Based Clustering Algorithm (MWBCA) [8]: MWBCA is designed as a reactive clustering algorithm based on the famous hierarchical protocol LEACH. It uses weighting function based on the evaluation of a score function called combined weight for clustering. In fact, the sensor node with the lowest weight is elected CH in order to balance the energy consumption in the network but researches such as [9] show that MWBCA has a very poor scalability.

5 Conclusion and Future Work

The harsh hardware limitations of sensor nodes make routing the most critical task in WSNs. Cluster Based Routing Protocol is one of the most important types of routing protocols in WSNs that can reduce the energy consumption and extend the network lifetime. This paper surveys the routing protocols classes and mainly focuses on hierarchical based routing protocols in WSN. We have systematically analyzed these routing mechanisms in and presented their representative protocols.

Our future work will be dedicated for an in-depth analysis of each of these hierarchical routing models under different conditions and with several experimental parameters in order to provide a punctual evaluation hierarchical routing protocols.

References

1. Agrawal, D.P.: Routing and performance of regular WSNs. In: Embedded Sensor Systems, pp. 329–351. Springer, Singapore (2017). https://doi.org/10.1007/978-981-10-3038-3_15
2. Akkaya, K., Younis, M.: A survey on routing protocols for wireless sensor networks. Ad Hoc Netw. **3**(3), 325–349 (2005)
3. Bana, S., Baghla, S.: Wireless sensor network. Int. J. Eng. Sci. **1706**, 1706–1712 (2016)
4. Shariff, S., Ahammed, G.F.A.: Performance analysis of routing protocols and energy models in WSN using QualNet. Adarsh J. Inf. Technol. **6**(1), 1–5 (2017)
5. Sabor, N., Sasaki, S., Abo-Zahhad, M., Ahmed, S.M.: A comprehensive survey on hierarchical-based routing protocols for mobile wireless sensor networks: review, taxonomy, and future directions. Wirel. Commun. Mob. Comput. **2017**, 23 (2017). https://doi.org/10.1155/2017/2818542. Article ID 2818542
6. Gautam, N., Lee, W.-I., Pyun, J.-Y.: Track-sector clustering for energy efficient routing in wireless sensor networks. In: Ninth IEEE International Conference on Computer and Information Technology, 2009, CIT 2009, pp. 116–121. IEEE (2009)
7. Buttyán, L., Schaffer, P.: Position-based aggregator node election in wireless sensor networks. Int. J. Distrib. Sens. Netw. **6**(1), 679205 (2010)
8. Fan, Z., Jin, Z.: A multi-weight based clustering algorithm for wireless sensor networks. College of Computer Science Educational Software Guangzhou University (2012)
9. Yadav, A.K., Rana, P.: Cluster based routing schemes in wireless sensor networks: a comparative study. Int. J. Comput. Appl. **125**(13), 31–36 (2015)

Using the Internet of Things to Monitor Human and Animal Uses of Industrial Linear Features

Kan Luo[✉], Sara Saeedi, James Badger, and Steve Liang

Department of Geomatics Engineering, University of Calgary,
Calgary, Canada
kan.luo@ucalgary.ca

Abstract. The boreal forest ecosystem of Alberta is increasingly affected by the human development related to natural-resource extraction, pipelines, roads and seismic lines. To evaluate the efficiency of restoration treatments, this project aims at the monitoring of physical conditions and human/wildlife presence on a recovered seismic lines ecosystem. For monitoring human and animal uses of industrial linear features, an Internet of Things (IoT) prototype system is developed with open source hardware, and open interoperable IoT standard. We try to build an accurate and cost-effective tool to detect the impact of human footprint and monitor the habit of the wild animal. So we implemented four types of trail counter which is a low-cost, open-source measurement device that counts pedestrian, bikes, all-terrain vehicles (ATVs), cars, and animals on trails, paths, and sidewalks.

Keywords: Internet of Things · OGC SensorThings API
Wireless sensor network · Open standard · Open hardware

1 Introduction

The boreal forest ecosystem of Alberta is increasingly affected by the human footprint related to natural-resource extraction, pipelines, roads and seismic lines. In order to mitigate these effects, different teams are working together to track the amount of human footprint present in a given area. However, it is difficult to collect human/animal uses in boreal areas due to its remote location, traffic inconvenience, the cold weather in winter, and the uncertainty of human/animal movement patterns. There is lacking a way to streamline the process to collect, communicate, share, and analysis field data.

A geosensor web is an internet-based information and communication system connected to one or many sensor networks in the field. Online connectivity allows users to operate the network and query the physical world from anywhere and at any time. With on-going advances of in situ sensing and communication technologies, geosensor webs offer the potential to produce the data required to describe and understand environmental conditions at unprecedented quantity and quality. A geosensor web can offer continuous temporal observations (e.g., micro-climate and wildlife presence) that either cannot be directly measured by traditional remote-sensing

© Springer International Publishing AG, part of Springer Nature 2018
M. R. Luaces and F. Karimipour (Eds.): W2GIS 2018, LNCS 10819, pp. 85–89, 2018.
https://doi.org/10.1007/978-3-319-90053-7_9

technologies, or are too costly to obtain through conventional field surveys. The objective of this research is to design, develop, test, and demonstrate a low-cost geosensor web prototype that will collect ground biophysical data (e.g. animal presence) along seismic lines, and well sites.

2 Human/Animal Tracking Literature Review

Juang [1] designed the ZbraNet animal tracking collar which used peer-to-peer network to deliver GPS data. However, not enough solar energy is available for harvesting GPS data, and is designed for zbra only. Based on ZbraNet, Reinholds [2] proposed LynxNet tracking collars for wild animal monitoring. LynxNet provided a more compact and lightweight solution, and employed a wide modality of sensors in addition to GPS location, that also provide data about the surrounding environment and help to detect patterns of activities of the animal. But this system didn't last long and supported short communication range. Alison [3] successfully used GPS collars in 280 deployments on 13 species to support 24 studies. Nevertheless, they spent a lot of money in retrieving and refurbing collars. TRAFx [4] counting systems are used for recreation, land use and visitor studies, counting vehicle traffic, trail use, bikes and off-highway vehicles. These products are field-proven, stable, accurate counters. However, they must use special equipment to retrieve data from counters in the field.

Most human/animal tracking solution stored data locally, and can not deploy equipment on a large scale due to cost. Collecting data is technique difficult and expensive, because their coverage of monitoring is small and their sensing node can not adapt to the changing environment. Integrated data analysis is also not included in these solutions. Locked to a particular company, which makes the data sharing difficult with these solutions.

3 System Architecture

Aiming at solving the existing problems in current animal/human tracking with wireless sensor network, we utilize the open source hardware, software, and open standard to solve the vendor lock. Using low power wide area network to expand the coverage of sensing area. The cloud can bring intelligence to analysis field data and control the field nodes remotely. Front-end dashboard enables data sharing and decision making.

The sensing node has been designed to be a smart object, which collects environmental conditions. It will adopt the component-based design, which supports users to add or remove hardware modules as needed. Therefore, the sensing node will be able to adapt to different scenarios. The node will also have the communication module to support bi-directional communication between each node. The battery or solar power system can extend the lifetime of the device from several days to several weeks, so that they can sustain themselves with reducing human participation.

4 Implementation

The proposed architecture can be implemented a platform which is inexpensive, open, and able to integrate different kinds of sensors, communicate, share, and analysis sensor measures.

In summer 2016, eleven IoT sensor nodes were deployed in three different sites in a boreal forest environment. Real-time ambient environmental data were sent back from the remote area via 2.5G mobile networks to an Open Geospatial Consortium (OGC) SensorThings API cloud server [5].

Last year, we implement four different kinds of trail counters, includes contact and contactless solutions, to understand how people or animals behave.

4.1 Trail Counter

The trail counter is a low-cost, open-source measurement device that counts pedestrian, bikes, ATVs and cars on trails, and paths. Trail counters are typically expensive devices used in the traffic monitoring to understand how people behave. Thanks to LinkIt Smart and open-source hardware, we're able to build a functional trail counter, includes contact and contactless solutions, to understand how human or animals behave quickly (Table 1).

Table 1. Components list included in four trail counters

Component	Description	Cost
Microcontroller board	LinkIt Smart 7688 Duo	$15.9
Breakout	Grove Breakout for LinkIt Smart 7688 Duo	$5.9
Ultrasonic ranger	Grove - Ultrasonic Ranger	$3.9
Passive Infrared Motion (PIR) sensor	Grove-PIR Sensor	$7.9
Differential pressure sensor	MPX5100GP Pressure sensor	$15.2
Acceleration sensor	Grove - 3-Axis Digital Accelerometer (±16 g)	$9.9

Contactless Solutions

The contactless solutions are based on ultrasonic ranger and PIR motion sensor. The principle of these two solutions are as followers: (1) If the ultrasonic ranger returns a distance that is short than observation thread hold or PIR sensor detect infrared changes which means the device observes a person or bike. (2) After the counter already recognized a person and will count number of times the person stays in front of the device. (3) If the counter hasn't seen the "person" for too long, and the count number of times the person stays in front of our device is more than minimum count to accept a person. The device will count one more person.

The advantage of contactless solution is the performance is not influenced by the daylight. The disadvantages are these counters cannot detect the direction of passing people or animals, and its max reliably covered distance is 2 m.

Contact Solutions

The contactless solutions are based on differential pressure sensor and acceleration sensor. The principle of differential pressure sensor solution: One end of rubber tube will connect to the upper lead of the pressure sensor. The other end of the tube must be hermetically sealed. This can be achieved by using an end plug or melting it. When the rubber tube is hit by a bike, car or an ATV, the air pressure inside the tube will change. We can identify the type of moving object based on the max pressure it generates, and estimate the speed of the moving object based on the travel time and average spacing between wheels of bike, car or ATV. The principle of acceleration sensor solution is quite similar with differential pressure sensor one. But the vibration sensor is installed on a board. The moving object will contact directly with the board instead of tube.

The contact solution can identify the type of moving object, and collect speed information. But it counts only through contacting, and cannot detect direction of passing bikes, cars or ATVs.

5 Conclusion

In this project, we design a sensor-network system based on existing commercial off-the-shelf open-source hardware, software architectures, and open standard.

By running the tests, we have gained the following insights from our preliminary results:

- Thanks to open hardware and IoT standard, we're able to build a functional, low-cost trail counter, and it's possible to large-scale deployment of this trail counter.
- The trail counter we try to build compares to the commercial trail counters, should be a cost-effective solution to detect the impact of human footprint and monitor the habit of them.

Our project is still in progress, many problems still need to be solved, but we firmly believe that open-source hardware, and interoperable IoT standard, and geospatial technologies could help us monitor boreal ecosystem very well.

References

1. Juang, P., Oki, H., Wang, Y., et al.: Energy-efficient computing for wildlife tracking: design tradeoffs and early experiences with ZebraNet. In: Proceedings of 10th International Conference Architecture Support for Programming Languages and Operating Systems (ASPLOS 2002), pp. 96–107 (2002). https://doi.org/10.1145/605397.605408
2. Zviedris, R., Elsts, A., Strazdins, G.: LynxNet: Wild Animal Monitoring Using Sensor Networks, pp. 2–5 (2009)

3. Matthews, A., Ruykys, L., Ellis, B., et al.: The success of GPS collar deployments on mammals in Australia. Aust. Mammal **35**, 65–83 (2013). https://doi.org/10.1071/AM12021

4. TRAFx Homepage. https://www.trafx.net/

5. Luo, K., Saeedi, S., Badger, J., Liang, S.: An OGC open standard-based Internet of Things prototype of vegetation recovery monitoring in Northern Alberta background and relevance. In: Spatial Knowledge and Information Canada 2017, Banff, Canada (2017)

Multigranular Spatio-Temporal Exploration: An Application to On-Street Parking Data

Camilla Robino[1], Laura Di Rocco[1], Sergio Di Martino[2(✉)] (iD),
Giovanna Guerrini[1] (iD), and Michela Bertolotto[3]

[1] University of Genova, Genoa, Italy
S3707993@studenti.unige.it, laura.dirocco@dibris.unige.it,
giovanna.guerrini@unige.it
[2] University of Napoli Federico II, Naples, Italy
sergio.dimartino@unina.it
[3] University College Dublin, Dublin, Ireland
michela.bertolotto@ucd.ie

Abstract. Traffic congestions cost billions of dollars to the society every year and are often aggravated by road users looking for parking. One way of alleviating the parking problem is providing decision makers of smart cities with powerful exploratory tools to analyse the data and find more effective solutions. This paper proposes a novel visual analytics tool for decision makers that allows multigranular spatio-temporal on-street parking data exploration. Even if the tool has been designed to deal with on-street parking data, it relies on a generic logic that makes it adaptable to more general spatio-temporal datasets.

Keywords: Data visualization · Intelligent Transportation Systems
Spatio-temporal analysis · Exploratory geo interfaces · Smart cities

1 Introduction

Traffic congestions cost billions of dollars to the society every year and represent a key stress factor for road users [1]. It is well known that in urban scenarios a significant fraction of traffic is caused by drivers looking for parking. A research conducted by Shoup in 2006 in Los Angeles (USA), showed that during rush hour, about 30% of the traffic was due to drivers looking for a parking space [2].

To face these problems, a lot of research in the field of Intelligent Transportation Systems (ITS) is focused on IT solutions to exploit the existing road infrastructure in a smarter way. Among them, a key topic is designing powerful exploratory tools for decision makers of smart cities [3,4]. These specifically tailored *data-driven* ITS are needed to interactively explore massive amounts of mobility data that are collected daily, allowing thus the decision makers to acquire a deeper insight on the mobility phenomena on an urban scale [5].

© Springer International Publishing AG, part of Springer Nature 2018
M. R. Luaces and F. Karimipour (Eds.): W2GIS 2018, LNCS 10819, pp. 90–100, 2018.
https://doi.org/10.1007/978-3-319-90053-7_10

For this kind of applications, data visualization techniques are widely recognized as powerful and effective [3,6], as they can fully exploit human abilities to perceive visual patterns and to interpret them [7,8].

The majority of these exploratory ITS solutions are aimed at supporting the analysis of traffic data and vehicular trajectories [3], while a surprisingly low amount of research efforts has been devoted to the definition of visual analytic tools for parking data [4,9]. Indeed, while there is a proliferation of apps to help users looking for a parking find their best (closest, cheapest,etc.) available option by analysing real time data[1] [10], there is a lack of applications to support decision makers by providing visualization, aggregation and analysis functionality for data collected over extended periods of time. It is worth noting that the monitored parking stalls have a fixed position in space, therefore, differently from trajectories, the collected data (i.e., parking availability) change only with respect to time, thus requiring different exploratory logic and visualization metaphors. To fill this gap, in this paper, we address the issue of on-street parking data exploration for decision makers (e.g., local authorities of a smart city). In particular, we define a logic for multigranular spatio-temporal exploratory analysis, and design a corresponding Graphical User Interface (GUI) that could be used by decision makers when investigating different parking strategies and plans in a city. The interface provides the functionality needed to analyse massive amounts of parking availability data, collected over an extended period of time, supporting both the aggregation of data at different levels of granularity, and multiple data comparisons, to highlight specific trends. In our work, the focus is on analytical tasks rather than on real time visualization. Therefore the main queries of interest are aggregations of the subset of data that satisfy the criteria specified by the user via the interface as well as comparison queries based on both spatial and temporal criteria. More in details, spatial multigranularity is achieved by allowing the user to zoom in and out on a map, with consequent recalculation of the corresponding visualization results. As for multiple temporal granularities, a specific temporal query panel is included in the interface, where the user can select to view data by month, day, hour etc. In order to understand the impact the proximity to specific points of interest (e.g., services) might have on data distribution, our interface also allows the visualization of points of interest on the map.

The main contribution of our work includes the formalization of a generic logic for the analysis and visualization of spatio-temporal discrete data, with focus on multigranular queries for aggregating properties of the data. This logic, independent of the specific domain the data refers to, is used for the definition of a specific visual analytics tool, for interactive exploration of on-street parking data aimed at decision makers of smart cities.

The remainder of the paper is structured as follows: Sect. 2 presents state of the art on parking data visualization, Sect. 3 shows our parking data visualization

---

[1] https://ischool.uw.edu/capstone/projects/2015/uw-parking-hero, http://map.wisc. edu/.

application and Sect. 4 discusses an example of a decision-maker session. Section 5 concludes the article discussing future work.

2   Background and Related Work

A lot of research has been oriented towards the development of ITS solutions for smart parking [4]. Limited to data visualization for smart parking solutions, most of the proposals are part of Parking Guidance and Information (PGI) systems, meant to present parking space location and availability to a mobility user. The majority of current portable navigation devices and in-vehicle connected navigators offer this feature. Anyhow, to date PGIs support only off-street parking facilities, like parking garages, silos, etc.[2]. Indeed, collecting on-street parking space availability information is still a challenging issue [9,11], which can be addressed either with the deployment of stationary sensors (more accurate, but very expensive to install and maintain on an urban scale) [12,13], or by means of participatory or opportunistic crowd-sensing solutions [14] from mobile apps (very cheap, but they need extremely high penetration rates) [13,15,16] or probe vehicles [16,17] (promising compromise, leading to a significant spatio-temporal sensing coverage) [12]. A well-known proposal of GUI for on-street parking is the one developed within the *SFParkSummary2014* project of the San Francisco Municipality, ran between 2011 and 2014, and which costed in excess of $46 million [9]. In this project, more than 8,000 parking spaces were equipped with static sensors embedded in the asphalt, thus providing on-street parking data, while a web application was developed to display the occupancy of parking spots and the current price, in two different visualizations, using scales of colours. However, the emphasis in such a system was on availability-dependent parking prices, by changing the parking tariff based on the availability of parking in the target area, to re-direct drivers towards areas with lower occupancy.

As for solutions targeted at decision makers of smart cities, ParkItSmart [18] is a web application to monitor specific parking areas in Zurich, as well as an application for drivers to choose parking based on availability of pay parking, free parking, off-street parking, special needs parking, etc.

Some remarkable visual analytics solutions have been developed for traffic. For example, *Ferreira et al.* [6] developed a system for visually querying taxi trips in New York City[3]. This model is able to express a wide range of spatio-temporal queries, and it allows the user to compose and aggregate queries and then visualize, explore and compare results. Interactive response times are achieved making use of an adaptive level-of-detail rendering strategy to generate clutter-free visualization for large results. The Dublin Dashboard[4] groups data from different data sources providing interactive data visualization about real-time information, time-series indicator data and interactive maps about different aspects of the city, including a visualization of off-street parking spaces.

---

[2] https://ischool.uw.edu/capstone/projects/2015/uw-parking-hero, http://map.wisc.edu/.

[3] see TaxiVis project http://vgc.poly.edu/projects/taxivis/.

[4] http://www.dublindashboard.ie/.

3   The Proposed Visual Analytics System

In this section, we present our approach for the multigranular visualization and analysis of spatio-temporal data and we propose a GUI for decision makers.

3.1   Data and Design Requirements

We visualize parking availability according to its spatio-temporal dimensions. While in many off-street parking applications availability is intended as the absolute number of free parking spaces (e.g., 3 free spaces at level 2 in car park 5), in the case on-street parking, the total number of spaces in a given street segment is usually quite low. Thus a better measure of availability is given by the ratio between free parking spaces and total number of parking spaces, particularly when analysing and aggregating data collected over an extended period of time. Therefore in this paper:

$$Availability = \frac{\#(free\ parking\ spaces)}{\#(parking\ spaces)}$$

We focus on availability at different temporal granularities showing data on a map as well as a chart trend. Given the target, the interface visualizes historical parking availability data, to support the decision maker in answering questions like: *"How does parking availability change over the Mondays in March?"* or *"How does the availability near shopping malls change in the two days before Christmas?"* but also *"How does parking garage availability change during the weekends?"*. Our spatio-temporal parking viewer will help a decision maker answer this kind of questions in an effective visual way.

3.2   The Proposed GUI

The application consists of two linked interfaces: the homepage and the comparative page. The homepage, shown in Fig. 1, is divided into two main panels:

- the *Availability Visualization Panel*, containing a map and a bar chart representing parking availability at different granularities (on the left hand side),
- the *Querying/Filtering Panel*, containing a timeline representing different granularity levels, parking facilities filtering and points of interest filtering (on the right hand side).

Availability Visualization Panel. The main component in this panel is a map of the area of interest, showing the target area under monitoring. The on-street parking availability is represented in two different ways on the map. A *heat map* is used when a large area is shown. When the area becomes smaller (by zooming in), we represent availability on the streets by *coloring the segments* in different colors. A *red-blue-light blue* scale is used, where the red nuance indicates

Fig. 1. Application homepage. (Color figure online)

lower availability and the light blue represents higher availability, similarly to the SFPark project [9].

The lower part of the panel contains a bar chart, which visualizes the average availability, with standard deviation, aggregated across all the segments shown in the map. We use the same color scale for bar chart as for the map, to immediately provide a visual feedback on the parking availability. The temporal granularity of the data shown on the map and with the bar charts is dictated by the corresponding Querying/Filtering Panel, described in the following.

The decision maker can zoom to a specific location of his/her interest in two different ways: by directly interacting on the map or by searching for specific places, using a search bar on the top left. As a decision maker scrolls/zooms on the map, the parking availability for the corresponding represented area is calculated and visualized, by aggregating data corresponding to all the displayed street segments. This functionality enables multigranular spatial analysis. The availability is calculated by a window spatial query.

Querying and Filtering Panel. The right side of the GUI contains a set of controls suited to select data to be shown in the *Visualization Panel*. More specifically, moving downward, the panel allows the user to: (*i*) perform temporal filtering; (*ii*) choose among different attributes of parking facilities; (*iii*) enable/disable the representation of Points of Interests (POIs) on the map; (*iv*) enable an overlay on the map of another datasource. The panel also contains a button *Query comparison* that links to the Comparative Page (see Fig. 2). The parking facilities filtering allows to analyse different types of parking. Consequently the map changes, showing only selected data. Given the parking classification, it is possible to analyse particular behaviours. As for the POIs, the decision maker can decide to add on the map different POIs, by selecting from categories and subcategories. In order to implement this, we connect our dataset to OpenStreetMap[5] (OSM) data. Specific points of interest shown on the map give the opportunity to understand, for example, relationships between parking availability and specific buildings and services.

Comparative Page. In the comparative page in Fig. 2 the decision maker can compare the results of two different queries. The page contains two identical panels where the visualization and temporal query panels are positioned one above the other. The panel semantics and functionalities (except for the filtering and linking to the other datasets) are the same as for the home page. On the top, there is the temporal query panel where the decision maker can choose the temporal granularity. The central part contains a visualization map that represents parking availability following the rules in Table 2. On the bottom, a bar chart is displayed that summarizes availability trends according to the rules in Table 2. Notice that, due to the limitations of space availability on the screen, in the comparative page, only temporal filters can be applied.

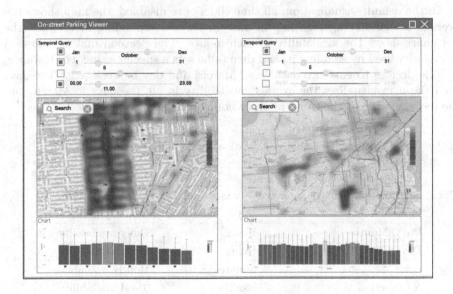

Fig. 2. Application comparative page.

3.3 The Logic for Spatio-Temporal Filtering

Assuming we have data spanning over a calendar year, we consider four different temporal granularities: *Month* (M), *DayOfMonth* (Dm), *DayOfWeek* (Dw) and

---

[5] http://www.openstreetmap.org/.

Hour (H). For each granularity, the user can choose a specific value by moving the slider. All these query bars can be enabled/disabled (E/D). If the user enables the temporal granularity *DayOfWeek*, the system automatically forbids to enable the *DayOfMonth* line and vice versa. This is because if, for instance, a day of week (e.g., Tuesday) is chosen, a day of month cannot be specified (e.g., 3rd) as the selected day might not correspond to a Tuesday. Table 1 shows the rules related to the combined selection of different temporal parameters. When a slider is marked with "E/D" this is means that the two selections are mutually exclusive.

Table 1. Rules for temporal sliders Enabling/Disabling (E/D).

Enable/Disable	Month	DayOfMonth	DayOfWeek	Hour
Month		E/D	D/E	E
DayOfMonth	E		D	E
DayOfWeek	E	D		E
Hour	E	E/D	D/E	

In the default visualization, all time sliders are disabled: the map shows the average availability aggregating all data in the corresponding area, while the bar chart shows the availability in each month. The decision maker chooses a granularity level by enabling it, and then (s)he sets a specific value. For instance, if (s)he decides to enable month (e.g., March) and day of week (e.g., Tuesday) then the map visualizes the average availability in all Tuesdays in March, and the bar chart shows the trend in every hour of the day, as in Table 2.

Table 2. Dynamic visualization depending on granularity.

Dynamic visualization		
Selected temporal granularity	Charts granularity	Map visualization
Month	Days	Monthly availability
DayOfMonth	Hours	Daily availability
DayOfWeek	Hours	Daily availability
Hour	Minutes	Hourly availability
No selection	Months	Total availability

The range of the x axis in the bar chart depends on the temporal query. Table 2 shows the relationships among the smallest temporal granularity selected, the x axis granularity in the chart, and the corresponding map visualization.

As for the aggregation method chosen for visualizing data in the chart and on the map, it depends on the area rendered on the map and on the selected temporal query. In Table 3, we show all the possible temporal queries that a decision maker can submit (depending on the mandatory behaviour of the temporal sliders) and highlight how the aggregation method changes. Notice that, when the selected granularity is *Hour*, we retrieve and then show the original data from the sensors. In all the other cases, we perform a preliminary aggregation phase of the data matching the spatio-temporal query, and then depict the average and the standard deviation of these aggregated data.

Table 3. Aggregation method depending on a query. D* represent either DayOfMonth or DayOfWeek. These two granularities have the same behavior.

Selected temporal query	Aggregation query
M-D*-H	`Availability WHERE M=m AND D*=d* AND H=h`
M-D*	`avg(Availability), sd(Availability) WHERE M=m AND D*=d* GROUP BY H`
M-H	`Availability WHERE M=m AND H=h`
M	`avg(Availability), sd(Availability) WHERE M=m GROUP BY Dm`
D*-H	`Availability WHERE D*=d* AND H=h`
D*	`avg(Availability), sd(Availability) WHERE D*=d* GROUP BY H`
H	`Availability WHERE H=h`
-	`avg(Availability), sd(Availability) GROUP BY M`

4 An Example of a Decision-Maker Session

Ideally, the decision maker starts by scrolling the map to the location of his/her interest. Then, (s)he selects the temporal granularity in the temporal query panel. For example (s)he enables the *Month* slider (e.g., March) and *DayOfWeek* slider (e.g., Tuesday). On the visualization panel, the map shows the average availability on all Tuesdays in March in the target area. The bar chart shows the trend of the average availability in all hours of the day. Each bar shows the standard deviation measure and the minimum and maximum availability in the corresponding hour. If the decision maker clicks on a specific bar, the map visualizes the availability in the selected hour. If (s)he clicks on minimum or maximum availability in a specific bar, the specific street is highlighted on the map. The user can enable or disable the temporal query and the map is updated with the new settings. Under the query panel, (s)he can choose to filter the map visualization with parking facilities. Furthermore, the user can decide to add on the map the points of interest by selecting them based on the available categories and subcategories, like for instance a Stadium.

At the end of this session, the decision maker decides to compare two different spatio-temporal queries. To do it, (s)he clicks on button *Query comparison* in the temporal query panel. A new page opens with the interface shown in (Fig. 2). The use of the GUI is the same as in the homepage. For example, if on the left the decision maker enables the *Month* slider (e.g., October), *DayOfMonth* slider (e.g.,5th) and *Hour* slider (e.g., 11.00 AM), on the visualization map, the system shows the availability on 5th March at 11.00 AM. Meanwhile the bar chart shows the trend of availability between 11.00 AM and 12.00 AM every 5 min. If on the right, the decision maker enables the *Month* slider (e.g., March) and the *DayOfMonth* slider (e.g., 5th), the corresponding map shows the average daily availability on 5th October. Meanwhile the bar chart shows the average availability trend every day of October. This way the parking availability on two different days can be compared.

5   Conclusions and Future Work

A wider diffusion of Intelligent Transportation Systems can ease mobility problems, by allowing mobility users and decision makers to use the existing road infrastructure in smarter ways. In this paper, we presented a visual analytics system targeted at decision makers of smart cities, meant to support them in planning mobility strategies by means of exploratory analysis of on-street parking data. In particular, a decision maker can analyse data according to both spatial and temporal dimensions, exploring combined visualizations of raw data on a map and also as aggregated data on a bar chart. The application provides the possibility to interact with the map in order to allow a decision maker to express and run different spatial analysis queries on the dataset. The temporal queries are handled by means of a specific multigranular filtering panel. An additional strength of our GUI is the possibility to show on the map (provided by OpenStreetMap) some Points of Interest, helping the decision maker to understand how the parking availability is influenced by surrounding phenomena. Moreover, the application includes a comparative page that allows two different queries to be visualized at the same time. For example, a decision maker can compare the results of queries over two different time periods on the same area or vice versa.

The main contributions of our data visualization application include:

- the design and development of a Graphical User Interface aimed at decision makers for on-street parking data analysis;
- a generic logic for the development of a system that enables storage, visualization and analysis of different spatio-temporal datasets.

Since the data that we manipulate relates to on-street parking, the interface is specifically modelled for this data, and as such it includes some specific filters for this case study (as shown in Sect. 3). Anyhow, our aim is to develop the back-end of our application as generic as possible, in order to allow other datasets with spatio-temporal dimensions to be used. The mandatory behavior that our visualization implements is strictly formalized on spatial queries and temporal queries. The spatio-temporal datasets that our application can analyse are those where it is not the spatial component that changes over time but some other property. An example could be the humidity levels recorded by sensor located at fixed positions. Some of the functionalities offered in our interface could also be useful for the general public (e.g., drivers looking for a parking, etc.). An interface for end-user will allow, for example, for remote booking of parking spaces and will include functionality for recommending the best parking plan that matches the user's criteria. We are also planning a mobile interface to help end-users look for the best parking solution. Given the different characteristics of a mobile screen, the interface will be based on the use of pop-up menus.

This work is at a preliminary stage. We are currently working on the full implementation of the system, also using D3.js[6], which is very useful for defining interactive pages on large datasets. Once a stable version of the system will be available, we will conduct an empirical evaluation of its effectiveness. To this aim we are already in touch with decision makers of the Mobility Agency of the city of Naples, Italy, that will be involved in this assessment.

References

1. Kwoczek, S., Di Martino, S., Nejdl, W.: Predicting and visualizing traffic congestion in the presence of planned special events. J. Vis. Lang. Comput. **25**(6), 973–980 (2014)
2. Shoup, D.: Cruising for parking. Transp. Policy **13**(6), 479–486 (2006)
3. Chen, W., Guo, F., Wang, F.Y.: A survey of traffic data visualization. IEEE Trans. Intell. Transp. Syst. **16**(6), 2970–2984 (2015)
4. Lin, T., Rivano, H., Le Mouël, F.: A survey of smart parking solutions. IEEE Trans. Intell. Transp. Syst. **18**, 3229–3253 (2017)
5. Zhang, J., Wang, F.Y., Wang, K., Lin, W.H., Xu, X., Chen, C.: Data-driven intelligent transportation systems: a survey. IEEE Trans. Intell. Transp. Syst. **12**(4), 1624–1639 (2011)
6. Ferreira, N., Poco, J., Vo, H.T., Freire, J., Silva, C.T.: Visual exploration of big spatio-temporal urban data: a study of New York city taxi trips. IEEE Trans. Vis. Comput. Graphics **19**(12), 2149–2158 (2013)
7. Compieta, P., Di Martino, S., Bertolotto, M., Ferrucci, F., Kechadi, T.: Exploratory spatio-temporal data mining and visualization. J. Vis. Lang. Comput. **18**(3), 255–279 (2007)
8. Andrienko, N., Andrienko, G., Gatalsky, P.: Exploratory spatio-temporal visualization: an analytical review. J. Vis. Lang. Comput. **14**(6), 503–541 (2003)
9. SFMTA: SFPark: Putting Theory Into Practice. Pilot project summary and lessons learned (2014) Accessed 24 June 2016
10. Di Napoli, C., Di Nocera, D., Rossi, S.: Agent negotiation for different needs in smart parking allocation. In: Demazeau, Y., Zambonelli, F., Corchado, J.M., Bajo, J. (eds.) PAAMS 2014. LNCS (LNAI), vol. 8473, pp. 98–109. Springer, Cham (2014). https://doi.org/10.1007/978-3-319-07551-8_9
11. Richter, F., Di Martino, S., Mattfeld, D.C.: Temporal and spatial clustering for a parking prediction service. In: IEEE 26th International Conference on Tools with Artificial Intelligence (ICTAI) 2014, pp. 278–282. IEEE (2014)
12. Bock, F., Attanasio, Y., Di Martino, S.: Spatio-temporal road coverage of probe vehicles: a case study on crowd-sensing of parking availability with taxis. In: Bregt, A., Sarjakoski, T., van Lammeren, R., Rip, F. (eds.) GIScience 2017. LNGC, pp. 165–184. Springer, Cham (2017). https://doi.org/10.1007/978-3-319-56759-4_10
13. Xu, B., Wolfson, O., Yang, J., Stenneth, L., Yu, P.S., Nelson, P.C.: Real-time street parking availability estimation. In: 14th International Conference on Mobile Data Management, vol. 1, pp. 16–25. IEEE (2013)

---

[6] https://d3js.org/.

14. Rinne, M., Törmä, S., Kratinov, D.: Mobile crowdsensing of parking space using geofencing and activity recognition. In: 10th ITS European Congress, pp. 16–19 (2014)
15. Ma, S., Wolfson, O., Xu, B.: Updetector: sensing parking/unparking activities using smartphones. In: ACM SIGSPATIAL International Workshop on Computational Transportation Science, pp. 76–85. ACM (2014)
16. Bock, F., Di Martino, S., Sester, M.: What are the potentialities of crowdsourcing for dynamic maps of on-street parking spaces? In: 9th ACM SIGSPATIAL International Workshop on Computational Transportation Science, pp. 19–24. ACM (2016)
17. Mathur, S., Jin, T., Kasturirangan, N., Chandrasekaran, J., Xue, W., Gruteser, M., Trappe, W.: ParkNet: drive-by sensing of road-side parking statistics. In: 8th International Conference on Mobile Systems, Applications, and Services, pp. 123–136. ACM (2010)
18. Tsiaras, C., Hobi, L., Hofstetter, F., Liniger, S., Stiller, B.: parkITsmart: minimization of cruising for parking. In: 24th International Conference on Computer Communication and Networks, pp. 1–8. IEEE (2015)

Region-Aware Route Planning

Sabine Storandt[✉]

Department of Computer Science, University of Würzburg, Würzburg, Germany
sabine.storandt@uni-wuerzburg.de

Abstract. We consider route planning queries in road or path networks which involve a user preference expressed in relation to a spatial region, as e.g. 'from Nanjing to Shanghai *along* Yangtze river' or 'from home to work *through* Central Park'. To answer such queries, we carefully define relevant subgraphs of the network for each region-of-interest and guide the route towards them. To extract these subgraphs, we need to solve several non-trivial geometric problems (as computing weak visibility regions), which require to interpret the embedded network both as a graph and as an arrangement of line segments. We describe a suitable preprocessing framework, taking the special structure of road networks into account to increase its performance. Our query answering algorithm then allows to trade detour length against time spent within or close to the desired region. Using acceleration techniques, region-aware routes can be planned efficiently even in networks with millions of edges, and also when considering large or complex regions.

1 Introduction

Conventional route planning services and navigation systems usually compute the shortest or quickest route from the source to the destination. Specifying individual preferences is often impossible or limited to certain yes/no decisions as e.g. the usage of toll routes. But sometimes the user already has an idea how the route should roughly look like. For example, certain landmarks or landscape elements might be attractive to drive or walk by, as well as using certain streets or going through specific parks, districts, cities or other kind of areas. Such preferences can easily be expressed in a query by naming the region-of-interest (ROI) and a suitable *spatial relation*, see Fig. 1 and the following examples:

- 'from San Francisco to Los Angeles *via* Highway 1' (2D region: street)
- 'from Konstanz to Karlsruhe *on* Rhein-Radweg' (2D region: bicycle trail)
- 'from Graz to Salzburg *through* Bad Ischl' (3D region: town)
- 'from Salt Lake City to Fargo *through* Montana' (3D region: state)
- 'from Chicago to Toronto *around* Lake Erie' (3D region: lake)
- 'from Nazca to Ecuador *along* the Peruvian coast' (2D region: border/coast)
- 'from Florenz to Rome *along* the Apennines' (3D region: mountains)

For *via/on/through* the main question is where to enter and exit the ROI, for *along/around* it is important to define which street/path sections qualify, and

© Springer International Publishing AG, part of Springer Nature 2018
M. R. Luaces and F. Karimipour (Eds.): W2GIS 2018, LNCS 10819, pp. 101–117, 2018.
https://doi.org/10.1007/978-3-319-90053-7_11

Fig. 1. Envisioned result for the query 'from Lisbon to Madrid *through* Sierra de San Pedro' (with the region-of-interest highlighted in red). (Color figure online)

which route choice is reasonable in the end. The listed example queries already hint that ROIs might differ significantly in their extend, shape and accessibility. Nevertheless, we develop a general framework that allows to answer all such region-aware queries.

1.1 Related Work

To guide the route manually, so called via points can be defined in GoogleMaps and other route planning services. The resulting route is then a concatenation of piecewise quickest paths from the source to via point 1, from via point 1 to via point 2, ..., from via point k to the destination. Approaches based on automatically selected via points are also used to provide alternative routes [1]. For region-aware route planning one could select via points within or close to the specified region. But a reasonable selection is non-trivial and the number of necessary via points might be quite high. Consider for example a query containing '*through* a <park>'. Then the selection of a single via point in the interior of the park or of a certain entry or exit might already lead to an unreasonable route (or exclude more reasonable ones). For queries containing '*along* a <river>', following the bends of the river (or the border of some other non-convex region) often results in non-quickest paths. Hence via points would have to be dense in order keep the resulting route close to the ROI as illustrated in Fig. 2.

In a sequenced route planning query [2], the user specifies the source and the destination as well as a list of point-of-interest classes as e.g. *gas station* or *supermarket* (instead of a particular gas station or supermarket which would be just via points), and the goal is to find the shortest route that visits at least one representative of each class on the way. A ROI could also be interpreted as the class of all the points it contains or are close-by. But this might result in a route which intersects the ROI in a single point only. And that certainly does not comply with the meaning of either *via*, *through* nor *along*, see Fig. 3 for an example.

Recently developed approaches for personalized route planning [3–5] allow to incorporate individual preferences in the objective function for each query. To specify the preference, the user has to insert a weight vector where each entry

Fig. 2. Example showing the shortcomings of the via point approach for the query '*along* Yangtze river'. Despite the selection of two via points (big white circles) close to the river, the route deviates from the river before, after and in between the via points and also becomes non confluent.

Fig. 3. The query 'from Salt Lake City to Fargo *through* Montana' certainly aims at a longer part of the route being actually inside Montana (red) than the depicted route. (Color figure online)

reflects the importance of a certain route metric to him (as e.g. gas consumption, travel time or scenic landscape). The route returned is the one that minimizes the cost of the linear combination of all metrics with the user's weights serving as coefficients. But as these weights are set globally, local detours over ROIs can not be modeled.

Other methods custom-tailored to produce nice routes were considered in the context of jogging routes [6] or walks in forest areas [7]. Here, nice landscape or landmarks serve as attractors, making a route more likely to get close. But in region-aware route planning, one does not want a generally nice route but the 'attractor' is fixed by the user and the interaction with the ROI is fixed by the selected spatial relation. Hence previous approaches don't easily apply here.

Forbidding the usage of certain streets is usually achieved by assigning penalties to the respective edge weights. Penalties were also shown to be a useful tool for producing alternative route sets [8,9]. Flexible queries, in which the set of allowed edges can be restricted in different ways, were also studied in [10].

We will mainly focus on the 'positive' case, where entering or passing by a certain ROI is desired instead of avoiding it.

1.2 Road Network and Region Model

Throughout the paper, we assume a road (or path) network to be given as a weighted, undirected[1], connected graph $G(V, E)$, embedded in the plane. The embedding is fixed by the coordinates of the nodes (latitude and longitude), and an edge $e = (v, w)$ is a straight line interpolation between v and w. Therefore, we will use the terms edges and line segments interchangeably.

The region-of-interest can be an arbitrary simple polygon (possibly with holes). We always assume the boundary of the polygon to be given as a sequence of node coordinates which then induce a chain of straight line segments.

1.3 Contribution

We discuss which region-aware queries are meaningful and how a reasonable interpretation of the respective spatial relations should look like. We identify as an important step for query answering the extraction of the interior graph of a ROI (containing the part of the road network inside the ROI), and the circumference graph (containing the part of the road network surrounding the ROI). The interior graph will be important for answering *via* and *through* queries, the circumference graph for queries with the spatial relation *along* or *around*. We describe how these subgraphs can be obtained and stored in an efficient manner for all scales of ROIs. For query answering, we describe a method to determine sensible nodes to enter and exit the respective ROI subgraph in dependency of the source and destination location. Subsequently, we sketch engineering techniques which allow to reduce the query time significantly. We evaluate our approach on road networks and ROIs extracted from Open Street Map, as well as artificial ones. Using a variety of example queries, we empirically prove the efficiency and meaningfulness of our approach.

2 Query Answering Pipeline

Despite the many different kinds of possible region-aware queries, we identify a common query answering pipeline. Our generic query template looks like this: 'from <A> to <B> <spatial relation> <ROI>' where A and B are placeholders for the start and destination (either given by address, POI name or coordinates), the spatial relation is among *via/through* or *along/around*, and the region is specified by name or as an arbitrary input polygon.

---

[1] Directed networks can also be handled with minor modifications. We restrict ourselves to undirected graphs here for a cleaner exposition of the algorithms.

1. **Subgraph extraction.** For the given ROI, we first identify the subgraph of the road network containing the edges that are relevant for the region w.r.t. the specified spatial relation, so the road set either qualifying as driving *through* or *along* the ROI. In addition, we determine suitable nodes to enter and exit this subgraph.
2. **Computing a set of reasonable routes.** For planning the route, we assume that the user does not want to drive unnecessary detours. So the partial routes from the source towards the ROI subgraph and from the ROI towards the destination should be shortest paths. Within the ROI graph, we will also restrict us to shortest paths. Then the route is well-defined by the specification of a ROI graph entry and exit point. Hence it remains to find the best entry/exit point pair. We require them to be not too close to each other (as otherwise the time spent within the ROI is too short, as e.g. depicted in Fig. 3). The remaining entry/exit combinations induce then a set of route candidates.
3. **Final selection.** From the resulting set of routes, we select one (or a few) to be presented to the user. We will use the *gain* of each route to make that decision, where the gain weighs time spent within the ROI graph against detour length.

If the region is known to the data base, steps 1 and 2 can be preprocessed. In the following, we provide the details for each of the steps. We first focus on the efficient extraction of the relevant subgraphs of the road network for a region.

3   Interior Graphs of Regions

For a 2D region directly referencing a street or trail, the interior graph is just the region itself. For 3D regions, we define the interior graph as follows.

Definition 1 (Interior Graph and Entry/Exit Points). *Given an embedded graph $G(V, E)$ and a non-degenerated polygon P, the interior graph of P contains all edges of G which intersect P. For the edges that intersect the boundary of P, we call their endpoints outside P the entry/exit points of P.*

For the efficient extraction of the interior graph, we assume that a grid data structure is available where each edge of the graph is indexed in every grid cell it intersects. Choosing the grid cell length as a multiple of the maximum edge length, it can be assured that each edge intersects at most three grid cells.

3.1   Computing Entry/Exit Points

For a given polygon P, we first compute all intersection points of edges with the boundary of P. So for each line segment of P we determine the grid cells which contain parts of this segment, and then check for intersection with all edges indexed in these cells. For each edge which intersects a boundary segment, it remains to determine if the endpoints lie within or outside the polygon. Note that also both endpoints can be outside or both inside (as shown in Fig. 4). For final

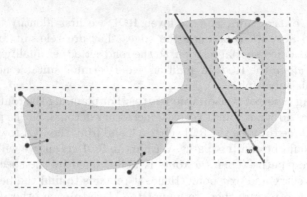

Fig. 4. Polygon (red) and all grid cells containing boundary segments (dashed squares). A few example edges (darkgray) intersecting the polygon boundary are drawn. For the edge $\{v, w\}$, with a single intersection point, the two possible elongating rays are given in blue and purple respectively. As the blue ray (over v) intersects 5 boundary grid cells and the purple ray (over w) only 1, we inspect the purple ray to decide which endpoint is inside. As the purple ray has 0 intersection points with the polygon boundary, we deduce correctly that w is outside and v inside, and hence w is an entry/exit point. The entry/exit points of all edges are indicated by red circles. (Color figure online)

assignment of inside/outside labels, we use ray casting. In ray casting, one emits a ray in arbitrary direction from a point, and the number of intersection points of the ray with the polygon boundary reveals whether the point is contained in P or not. In our scenario, we can detect the labels for both endpoints v, w of an edge at once by defining the ray as an elongation of the edge over one of the endpoints. As we are always free to choose which of the two possible rays we use, we decide for the one which intersects the smallest number of boundary grid cells (see also Fig. 4).

For a typical road network and a benign polygon shape, the number of edges per grid cell as well as the number of boundary grid cells intersected by a line segment or ray is constant. If this applies, the computation of the entry/exit points takes time linear in the size of the polygon boundary.

3.2   Extracting Interior Edges and Skeletons

Once we computed all entry/exit points and know the set of edges crossing the boundary of P, we can compute the complete interior graph of P by just adding the set of edges contained in grid cells that are completely inside the polygon. Note that these grid cells are easy to identify once all boundary grid cells were marked. But for large ROIs (like states or even countries), storing the complete interior graph is too space-consuming to be practical. Also, when just wanting to drive *through* such a ROI, many of the interior edges are actually not important, as they are not part of any reasonable path from some entry point to some exit point of the ROI. Hence the path skeleton, as defined in the following, is a more concise way to store the important information.

Definition 2 (Path Skeleton). *Given a graph $H(V, E)$ and a set of nodes $S \subseteq V$, the path skeleton of (H, S) is a subgraph of H containing all nodes/edges on shortest paths between nodes in S.*

With H being the interior graph of the polygon, and S the set of entry/exit points, the path skeleton contains all nodes and edges necessary for route planning inside the ROI. For efficient storage, all chains of nodes of degree 2 can be stored as single edges with the edge weight being equal to the respective path cost, see Fig. 5 for an example of such a compressed skeleton.

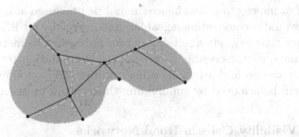

Fig. 5. Polygon/region with its interior graph (gray). The path skeleton between the entry/exit points is given in yellow, the compressed skeleton – containing significantly fewer nodes and edges – in green. (Color figure online)

4 Circumference Graphs of Regions

While the definition of the interior graph of a region is straight forward, specifying the subgraph relevant for *along*-queries becomes more involved. As a rationale, we assume that the user wants to see the ROI while driving along its boundary and that other edges in the network block the view. So for some edge $e \in E$, a polygon P is only visible from e if we can draw a straight line (of sight) from some point on the edge to some point on the boundary of P with the line not intersecting any other edges. Speaking in terms of computational geometry, the edge and the polygon have to be weakly visible from each other. Hence, we derive the following definition of the circumference graph of a ROI.

Definition 3 (Circumference Graph). *Given an embedded graph $G(V, E)$ and a polygon P, the circumference graph of P contains all edges of G which are weakly visible from the boundary of P (but do not intersect P).*

In an arrangement of k line segments, the problem of determining for two segments whether they are weakly visible from each other was proven to be 3SUM-hard and hence the time complexity of its solutions is $\Omega(k^2)$. An asymptotically optimal $\mathcal{O}(k^2)$ algorithm was presented in [11]. In our scenario, k is the number of edges m in the network plus the number of segments defining the boundary of the polygon, hence $k > m$. An algorithm with quadratic runtime is impractical here, as m tends to be in the order of millions for country-sized road networks. This also renders algorithms with a preprocessing time of $\mathcal{O}(k^{2+\epsilon})$ and subsequent query times of $\mathcal{O}(k^{1+\epsilon})$ as designed in [12] useless in our scenario.

For arrangements where the number of segment pairs which are indeed weakly visible from each other is low, output sensitive algorithms potentially perform better. In fact, the so called full visibility graph of an arrangement of k line segments (where each segment is a node in the graph and edges indicate weak visibility between segments) can be computed in time $\mathcal{O}(k \log k + B)$ where B is the resulting number of edges [13]. For $B \ll k^2$, the computation of the circumference graph of a single region in the preprocessing phase would certainly be possible. But with hundreds or thousands of possible ROIs (which can be nested or intersecting, and hence can not be processed at once), the preprocessing times would become unmanageable – and especially for ROIs being only specified at query time (e.g. via a self-drawn user polygon) this method does not allow for real-time query answering. To reduce the computation time, we first seek for an efficient way to find a (small) supergraph of the circumference graph of a ROI. This will be achieved by subdividing the network in so called visibility cells.

4.1 Visibility Cells in Road Networks

We use the concept of *potential visibility* to later define visibility cells.

Definition 4 (Potentially visible). *In a line segment arrangement, two segments l, l' are potentially visible from each other if one can draw an arbitrarily shaped line from l to l' which does not intersect any other segments.*

While for (weak) visibility the line connecting l and l' has to be straight, potential visibility does not have this requirement.

Definition 5 (Visibility Cell). *Given an embedded graph $G(V, E)$, we call a maximal subset of the edges $C \subseteq E$ which are all pairwise potentially visible from each other a visibility cell of G.*

Now our goal is it to subdivide G into all its visibility cells (see Fig. 6 for an illustration). If we assume the graph embedding is planar, the problem of computing the visibility cells is the same as reporting all faces of the embedding. Though real-world road networks contain a few edge crossings when embedded in the plane (as e.g. due to bridges), they can easily be planarized (by inserting additional nodes) for the sake of visibility cell computation.

If the maximum node degree of the network $G(V, E)$ is constant, all faces of the embedding can be listed in time linear in $|E|$ (see e.g. [14]): Each edge $\{v, w\}$ can be part of at most two faces, where the 'left' face can be found by walking counterclockwise around the face starting at w (always selecting the leftmost edge at the head node to continue), and the 'right' face analogously by going clockwise until v is reached. So computing a face or cell C of boundary complexity $|C|$ takes time $\mathcal{O}(|C|)$ and as we know that $\sum_C |C| \leq 2|E|$ the linear total runtime follows.

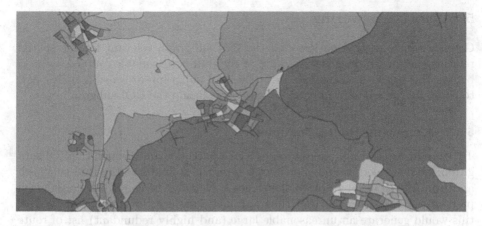

Fig. 6. Visibility cells indicated by different colors. (Color figure online)

With a subdivision into visibility cells, given any ROI, its circumference graph can only consist of edges in visibility cells that enclose parts of the ROI (but are not completely contained in the ROI). As visibility cells in road networks will turn out to be small and of low boundary complexity on average, this will lead to a significantly reduced runtime for extracting circumference graphs.

4.2   Weak Visibility Inside Cells

We augment our grid data structure such that every grid cell is indexed with the visibility cells that have a non-empty intersection (which can be achieved by treating the grid as a graph and using the methods for interior graph extraction for every visibility cell considered as ROI in the way described above).

For a given ROI, the set of intersecting visibility cells can the be aggregated by collecting the visibility cell indices within all its intersecting grid cells, and then testing for real intersection (via ray casting from all ROI boundary points). From the remaining visibility cells, we subtract the interior graph of the ROI. As a result we get a graph $G_S(V_S, E_S)$ where E_S describes the boundary of the ROI as well as all edges potentially visible from the ROI (but not inside the ROI). V_S is the respective set of endpoints of edges in E_S. Now, we can indeed apply the output-sensitive algorithm with a runtime of $\mathcal{O}(k \log k + B)$ [13] (or the suboptimal but more practical algorithm presented in [15] with a runtime of $\mathcal{O}(B \log k)$), as $k = |E_S|$ is now significantly smaller than $|E|$.

Once the circumference graph is available, we can compute entry/exit points and path skeletons in the same way as for the interior graph. Here, an entry/exit is defined as a node which is incident to an edge in G that is not part of that circumference graph.

5 Query Answering

Given a source, a destination, and the ROI subgraph relevant for the spatial relation specified by the user, we want to compute a reasonable route. As argued before, this involves to enter and exit the ROI subgraph at sensible nodes, and to compute the route from entry to exit point inside the ROI subgraph.

5.1 Computing Route Candidates

To compute a set of meaningful routes which partially run through the ROI subgraph, we first need to find good routes from the source s to the ROI subgraph and from the ROI subgraph to the target t. Naively, we could just consider all pairs of nodes v, v' in the ROI subgraph and compute the route s, v, v', t. But this would generate an unreasonable large (and highly redundant) list of route candidates to choose from. As each route has to go over an entry/exit node of the ROI graph anyway, no routes are lost when restricting the choice of v, v' to entry/exit nodes. While the number of entry/exit nodes is moderate for small ROIs or ROIs in sparse areas, the circumference graph of a long river still exhibits hundreds or thousands of entry/exit nodes. Again, generating a quadratic number of route candidates upon this entry/exit node set is expensive and also leads to many unreasonable routes. So we would like to prune unnecessary entry/exit nodes. We achieve that by restricting us to *locally optimal touch nodes (LOT nodes)* as defined in the following.

Definition 6 (LOT nodes). *Given a graph $G(V, E)$, a subgraph $H(V', E')$ of G, and a vertex $v \in V$, we call $L \subseteq V'$ the set of locally optimal touch nodes of v w.r.t. H with $l \in L$ if $d_v(l) \leq d_v(w)$ for all neighbors w of l in H.*

Figure 7 illustrates the concept.

Fig. 7. River with its circumference graph (pink) and entry/exit nodes. The green edges indicate the relevant part of the shortest path tree from v that defines which of these nodes are LOT nodes. As the shortest paths to the turquoise marked nodes enter the circumference graph already before they reach these nodes, the nodes can't be LOT nodes. So only the nodes marked with purple squares would be considered for route candidate generation. (Color figure online)

Finally, for each LOT node l of s and each LOT node l' of t, we add the route s, l, l', t to our candidate set, where $s \rightarrow l$ is a shortest path in G, $l \rightarrow l'$ is a shortest path in the ROI subgraph, and $l' \rightarrow t$ is again a shortest path in G.

5.2   Route Selection

For each route in the candidate set, we compute the total distance and the distance within the ROI subgraph. We call a route *reasonable* if no other route in the candidate pool is shorter and at the same time has a longer route section within the ROI subgraph. So unreasonable (or dominated) routes are pruned immediately. Furthermore, we prune routes where the time spent within the ROI is too short. To define that cut off, we use the lower quantile. For each of the remaining routes, we compute the *gain*. Here, the gain is defined as the ratio of time spent within the ROI graph and the detour length +1. The detour length is the total distance of the route minus the shortest path distance from s to t (without having to visit the ROI). The +1 term prohibits an infinite gain in case the shortest path already intersects the ROI subgraph. We then select the route with the highest gain that is not self-intersecting (or the one with the highest gain when all route candidates are self-intersecting).

Note that the gain can also be used to judge the meaningfulness of the region-aware query. If the gain of all route candidates is very small, it means one has to drive a huge detour to get to the ROI while spending a comparatively small time within the ROI graph.

6   Acceleration

We rely on shortest path computations in the preprocessing as well as on query time: In the preprocessing, we compute the path skeleton of a ROI subgraph by investigating all shortest paths between the entry/exit nodes. On query time, we compute shortest paths from the source to the ROI entry nodes and from the ROI exit nodes to the target.

While Dijkstra's algorithm is feasible to perform all these computations, it's typically too slow in country sized networks. In order to speed up all these steps, we use conventional preprocessing-based acceleration tools as contraction hierarchies (CH) [16] and (batched) PHAST [17,18]. Shortest paths inside the ROI subgraphs, however, might not coincide with shortest paths in the whole network (as the real shortest path might leave the ROI subgraph). As we want to forbid this, we block all edges from entry/exit nodes which are not inside the ROI subgraph. Then, using techniques for edge restrictions [10] or for updating a CH [19], we can adapt CH such that ROI subgraph queries are also answered correctly.

7   Experimental Evaluation

We implemented all described steps for region-aware route planning in C++. Experiments were conducted on a single core of an Intel i5-4300U CPU with 1.90 GHz and 12 GB RAM.

7.1   Road Network Data

We used the road network of Germany, extracted from Open Street Map[2], for evaluation. The network contains 22,046,972 nodes and 23,328,628 (undirected) edges. Computing a contraction hierarchy on this graph takes about 3 min and results in an overlay graph with 39 million edges. For a randomly chosen source and target node, Dijkstra's algorithm takes about 4 s on average to compute the optimal path, the same query can be answered in 1 ms using CH. A one-to-all query with Dijkstra takes about 6 s, and about 200 ms with PHAST. The performance of RPHAST for one-to-many queries depends on the size of the target set. For up to 2^{14} targets the query time never exceeded 10 ms. In the following queries, we will always use the accelerated variants.

The computation of the visibility cells took less than a minute (including time for planarization, which added 87,442 edges). In total, the network induces 1,357,310 visibility cells with the number of edges on the boundary ranging from 3 to 542,951 (the outer face). The average cell has a boundary complexity of only 36. More than 99.17% of cells have a boundary complexity of less than 500. The largest face after the outer face has a boundary complexity of 11,900. So all finite faces of the embedded network are of rather low complexity (three to six orders of magnitude smaller than the total number of edges in the network).

7.2   Region Data

We extracted real regions from Open Street Map. We selected 10 examples of each of the following categories, trying to have a large diversity w.r.t. size, boundary complexity, and accessibility within each category: lake, river, mountain, street, border, city.

As real region data is naturally inhomogeneous, and therefore average values over sets of real regions might give little information, we furthermore considered artificial ROIs. We used randomly positioned rectangles (low boundary complexity) of different height and width, and polygons created on random points clouds within a certain bounding box (high border complexity), using the 2-opt heuristic [20]. These artificial ROIs also nicely reflect the way a user might define some ROI manually to guide the route. The random placement of the artificial ROIs puts them often in dense regions of the graph, while some real ROIs as lakes or mountains naturally are in sparse regions or induce sparse regions themselves. Therefore we consider the artificial ROIs as worst case inputs and also use them to show the limits of our approach.

7.3   Subgraph Extraction

For each ROI, we computed its interior and its circumference graph. For one example from each category of real ROIs, the important characteristics of those graphs are collected in Table 1. While a single example from each category is not representative, tendencies which we also observed on the complete testbed become visible:

---

[2] openstreetmap.org.

(1) For natural regions (lakes, rivers, mountains), the complexity of the interior graph is usually below the boundary complexity of the ROI. The complexity of the circumference graph is within two times the boundary complexity of the ROI. If we compute the skeletons for both graphs, their complexity in total was always *below* the boundary complexity of the ROI. That means, a data base which holds all ROI polygons can be augmented with the ROI skeletons such that the space consumption less than doubles – but now region-aware route planning through/along all ROIs is possible.

(2) For man-made regions, as e.g. cities, the complexity of the ROI graphs is often significantly larger than the boundary complexity of the ROI. But again, using compressed skeletons, the additional space consumption decreases significantly. For the city in Table 1, the compressed skeleton of the interior graph has a size of 998, the compressed skeleton of the circumference graph a size of 1,502.

For comparison, Table 2 lists the results for artificially generated ROIs of different shape, area and complexity. We observe that already for a square of 5×5 km, the interior and the circumference graphs contains on average over 1,000 nodes. So the ratio of subgraph size and boundary complexity is very high for all our rectangular ROIs. Computing the skeleton reduces this ratio for both graphs significantly: For the 5×5 instances, the compressed skeleton contains on average less than 200 nodes. For rectangles of larger area, the size of the interior graph scales proportionally. The size of the circumference graph (as well as the number of boundary visibility cells) grows slower, as it is proportional to the

Table 1. Characteristics of interior and circumference graphs for selected regions. The row $|b(P)|$ contains the boundary complexity. The number of nodes in the respective subgraphs is denotes by 'size', #EN gives the number of entry/exit nodes, #VC the number of visibility cells that intersect the boundary. The 'time' rows provide extraction times in milliseconds.

	Lake	River	Mountain	Street	Border	City
$|b(P)|$	231	12,395	2,312	1,956	16,720	1,649
Interior graph						
Size	0	241	1,144	1,956	-	76,199
#EN	0	198	76	65	-	740
Time	0 ms	17 ms	4 ms	1 ms	-	139 ms
Circumference graph						
Size	435	15,212	4,438	-	28,101	23,458
#VC	1	26	21	-	510	73
#EN	39	1,170	406	-	6,114	1,356
Time	77 ms	693 ms	43 ms	-	866 ms	550 ms

Table 2. Characteristics of interior and circumference graphs for artificial regions, averaged over 100 examples for each column. The row 'w × h' gives the extension of the base area in kilometers. The other rows have the same meaning as in Table 1.

	■	■	▬	▬▬	✻	✻
$|b(P)|$	4	4	4	4	5,000	5,000
w × h	5 × 5	25 × 25	5 × 10	5 × 50	5 × 5	25 × 25
			interior graph			
size	1,706	59,112	3,300	27,742	813	18,793
#EN	53	443	93	579	238	3,609
time	4 ms	90 ms	10 ms	42 ms	41 ms	752 ms
			circumference graph			
size	2,149	11,276	3,677	13,545	3,189	29,212
#VC	16.2	89.2	20.4	134.2	174.1	792.8
#EN	186	1023	290	1,296	612	6,593
time	51 ms	221 ms	65 ms	400 ms	437 ms	2.8 secs

circumference of the ROI. The higher boundary complexity, as investigated in the last two columns of Table 2, increases the size of the circumference graph as well as the number of entry/exit nodes for the interior and circumference graph significantly. With more visibility cells to check and more boundary segments to investigate, computing the weak visibility graph becomes expensive.

Averaged over the 60 real ROIs and 600 artificial ROIs we tested, the extraction time of the interior graph is 120 ms, and 510 ms for the circumference graph. Including skeleton computation, the average time to preprocess a single ROI was about 1 s.

7.4 Query Answering

Table 3 shows our experimental results for query answering on the artificial ROI testbed. We observe that the number of LOT nodes is significantly smaller than the number of entry/exit nodes. Hence the runtime improves when using LOT nodes. Averaged over all 120,000 (6 × 100 ROIs × 200) conducted queries, the time for query answering is 36 ms only.

To check if our pipeline produces sensible routes, we inspected the results of 120 example queries (10 for each considered ROI category, real + artificial) where the gain exceeded 0.25. While it is difficult to come up with objective measures for their quality, we observed that the produced routes were all confluent (no U-turns), and that by mere visual inspection of the map no obviously better route could be identified. Figure 8 provides example query results.

Table 3. Experimental results for region-aware queries, averaged over 100 random queries per ROI per column and spatial relation. #LOT gives the number of LOT nodes, #cand. the number of reasonable route candidates

	■	▮	■	▬	🦅	♻
			via/on/over/through			
#LOT	9	67	8	97	35	112
#cand.	7	32	2	25	18	57
time	16 ms	21 ms	15 ms	29 ms	18 ms	33 ms
			along/around			
#LOT	37	192	55	236	96	305
#cand.	44	257	86	268	411	1,492
time	19 ms	56 ms	26 ms	65 ms	29 ms	113 ms

Fig. 8. Example queries for *along* a lake (left image) and *through* an artificial ROI (right image): shortest path without the ROI (black), ROI (red), region-aware route (blue), visibility cell (gray), interior graph (yellow), and alternative paths from the source and target to their LOT nodes (green and orange). (Color figure online)

8   Conclusions and Future Work

We showed that there are sensible route planning queries which involve a region-of-interest. Especially for named regions, such queries can be specified very concisely – hence they also might be of interest for interactive navigation. Our pipeline allows for real-time answering of region-aware queries (even for large and complex regions) when region specific subgraphs are precomputed. The resulting routes seem to reflect the intention behind the query well, and appear to be sensible on the local and the global level.

There are many directions for improving or extending our approach. So far, we have not considered queries over multiple regions or with combined relations as e.g. '*along* the Anglo-Scottish border but staying *in* Scotland'. Also other definitions of the circumference graph could be tested. For further acceleration (especially when complex regions are defined only on query time) methods for simplifying the region – without affecting the interior/circumference graph – could be helpful.

References

1. Abraham, I., Delling, D., Goldberg, A.V., Werneck, R.F.: Alternative routes in road networks. J. Exp. Algorithm. **18**, 1.3:1.1–1.3:1.17 (2013)
2. Eisner, J., Funke, S.: Sequenced route queries: getting things done on the way back home. In: Proceedings of the 20th International Conference on Advances in Geographic Information Systems, pp. 502–505. ACM (2012)
3. Funke, S., Storandt, S.: Personalized route planning in road networks. In: Proceedings of the 23rd SIGSPATIAL International Conference on Advances in Geographic Information Systems, p. 45. ACM (2015)
4. Dibbelt, J., Strasser, B., Wagner, D.: Fast exact shortest path and distance queries on road networks with parametrized costs. In: Proceedings of the 23rd SIGSPATIAL International Conference on Advances in Geographic Information Systems, p. 66. ACM (2015)
5. Funke, S., Storandt, S.: Personal routes with high-dimensional costs and dynamic approximation guarantees. In: Proceedings of the 16th International Symposium on Experimental Algorithms
6. Gemsa, A., Pajor, T., Wagner, D., Zündorf, T.: Efficient computation of jogging routes. In: Bonifaci, V., Demetrescu, C., Marchetti-Spaccamela, A. (eds.) SEA 2013. LNCS, vol. 7933, pp. 272–283. Springer, Heidelberg (2013). https://doi.org/10.1007/978-3-642-38527-8_25
7. Bast, H., Sternisko, J., Storandt, S.: ForestMaps: a computational model and visualization for forest utilization. In: Pfoser, D., Li, K.-J. (eds.) W2GIS 2013. LNCS, vol. 8470, pp. 115–133. Springer, Heidelberg (2014). https://doi.org/10.1007/978-3-642-55334-9_8
8. Kobitzsch, M., Radermacher, M., Schieferdecker, D.: Evolution and evaluation of the penalty method for alternative graphs. In: ATMOS-13th Workshop on Algorithmic Approaches for Transportation Modelling, Optimization, and Systems-2013, vol. 33, pp. 94–107. Schloss Dagstuhl-Leibniz-Zentrum fuer Informatik (2013)
9. Luxen, D., Schieferdecker, D.: Candidate sets for alternative routes in road networks. J. Exp. Algorithm. (JEA) **19**, 2–7 (2015)
10. Geisberger, R., Rice, M.N., Sanders, P., Tsotras, V.J.: Route planning with flexible edge restrictions. J. Exp. Algorithm. (JEA) **17**, 1–2 (2012)
11. Gajentaan, A., Overmars, M.H.: On a class of $O(n^2)$ problems in computational geometry. Comput. Geom. **5**(3), 165–185 (1995)
12. Nouri, M., Zarei, A., Ghodsi, M.: Weak visibility of two objects in planar polygonal scenes. In: Gervasi, O., Gavrilova, M.L. (eds.) ICCSA 2007. LNCS, vol. 4705, pp. 68–81. Springer, Heidelberg (2007). https://doi.org/10.1007/978-3-540-74472-6_6
13. Ghosh, S.K., Mount, D.M.: An output-sensitive algorithm for computing visibility graphs. SIAM J. Comput. **20**(5), 888–910 (1991)
14. de Berg, M., van Kreveld, M., Overmars, M., Schwarzkopf, O.C.: Computational geometry. In: Computational Geometry, pp. 1–17. Springer, Heidelberg (2000). https://doi.org/10.1007/978-3-662-04245-8_1
15. Overmars, M.H., Welzl, E.: New methods for computing visibility graphs. In: Proceedings of the Fourth Annual Symposium on Computational Geometry, pp. 164–171. ACM (1988)

16. Geisberger, R., Sanders, P., Schultes, D., Delling, D.: Contraction hierarchies: faster and simpler hierarchical routing in road networks. In: McGeoch, C.C. (ed.) WEA 2008. LNCS, vol. 5038, pp. 319–333. Springer, Heidelberg (2008). https://doi.org/10.1007/978-3-540-68552-4_24
17. Delling, D., Goldberg, A.V., Werneck, R.F.: Faster batched shortest paths in road networks. In: OASIcs-OpenAccess Series in Informatics, vol. 20. Schloss Dagstuhl-Leibniz-Zentrum fuer Informatik (2011)
18. Delling, D., Goldberg, A.V., Nowatzyk, A., Werneck, R.F.: Phast: hardware-accelerated shortest path trees. J. Parallel Distrib. Comput. **73**(7), 940–952 (2013)
19. Geisberger, R., Sanders, P., Schultes, D., Vetter, C.: Exact routing in large road networks using contraction hierarchies. Transp. Sci. **46**(3), 388–404 (2012)
20. Heidi, M.: Heuristics for the generation of random polygons*. In: Canadian Conference on Computational Geometry, vol. 5, p. 38. McGill-Queen's Press-MQUP (1996)

A Web Interface for Exploiting Spatio-Temporal Heterogeneous Data

Ba-Huy Tran[1](✉), Christine Plumejeaud-Perreau[2], and Alain Bouju[1]

[1] L3i Laboratory, University of La Rochelle, La Rochelle, France
ba-huy.tran@univ-lr.fr
[2] LIENSs Laboratory, U.M.R. 7266, CNRS and University of La Rochelle,
La Rochelle, France

Abstract. Nowadays, in several domains, data can be available from a multitude of sources using many models. Because of their heterogeneity, it is difficult to retrieve, combine and exploit this information. We present in this paper a Web interface for exploiting spatio-temporal heterogeneous data. The system is based on semantic data integration process using a materialization approach. A mechanism is proposed to enriched the knowledge base with spatio-temporal rules and experts knowledge. The system can be applied for various scenarios, and we illustrate this with two of our projects.

Keywords: Spatio-temporal · Heterogeneous data · Data integration
Ontology

1 Introduction

The dramatic rise of data availability, either in local storages or online, has increased the difficulty of retrieving relevant information. In fact, data can be available from a multitude of sources using many models. These are often heterogeneous because they are designed independently by different designers. To exploit these heterogeneous data, a data integration process has to be realized first and foremost. Data integration is a process that aims to first combine data from different sources by eliminating conflicts between them and then providing users with a unified view of the data. In the traditional data integration approach, data sources are queried through a global data model [1]. However, in general, there is not a unique conventional data model that is capable of accessing all available data with the granularity required by all users. One of the benefits often stated for RDF is the ease with which data can be integrated from distributed RDF sources [2]. At the heart of semantic data integration is the concept of ontology which defines a conceptual representation of data and their relationships. An ontology is then used to smooth differences between models by mapping them to the shared concepts. Ontology acts as a mediator for re-conciliating the heterogeneities between different data sources [3].

© Springer International Publishing AG, part of Springer Nature 2018
M. R. Luaces and F. Karimipour (Eds.): W2GIS 2018, LNCS 10819, pp. 118–129, 2018.
https://doi.org/10.1007/978-3-319-90053-7_12

With the purpose of integrating spatio-temporal heterogeneous data by the mean of ontologies, spatio-temporal ontologies have been developed. They are used to describe data sources, especially the dynamic of objects and their spatio-temporal relations throughout time. In the literature, several spatio-temporal ontologies [4,5] have been introduced following the perdurantist approach. These models consider that objects have several time slices in their lives. During a time slice, the properties of an object are fixed, and the whole is valid during a certain interval of time or at a given set of instants.

This paper presents the architecture of a semantic information system having a Web interface for exploiting spatio-temporal heterogeneous data. The data sources are semantically integrated thanks to spatio-temporal ontologies to be exploited afterward. Section 2 exposes the theoretical background underlying a semantic data integration process. Section 3 examines current geospatial triplestores, especially their performances and the expressiveness of the queries for building a knowledge base having spatio-temporal reasoning capabilities. Sections 4 and 5 describes the system architecture and its application in two different use cases. Finally, Sect. 6 draws the conclusions.

2   Semantic Integration of Spatio-Temporal Data

Semantic integration system use ontologies in their integration process. Because of their potential for describing the semantics of knowledge, ontologies can play multiple roles in the integration process. [3] have identified five main roles of ontology in data integration process.

- Metadata representation: metadata in each data source can be explicitly represented by a local ontology. These ontologies are homogeneous since they use the same representation language;
- Global conceptualization: the global ontology provides a conceptual view of the heterogeneous source schemas;
- Support for high-level queries: ontology provides a high-level view of sources. Therefore, a query can be formulated without specific knowledge of data sources. Using an ontology as a query model, the structure of query model should be more intuitive for users because it corresponds more to their appreciation of the domain;
- Declarative mediation: the global ontology can be used as a mediator for query rewriting across peers;
- Mapping support: a common vocabulary, which can be formalized as an ontology, can be used to facilitate the mapping process. Since ontologies contain a complete specification of the conceptualization, the mappings can be validated with respect to ontologies to facilitate its automation.

The semantic data integration process can be divided into three stages: ontologies construction, ontologies integration, and data integration.

Ontologies Construction

The first stage aims to construct ontologies from sources schema. This process is realized manually or semi-automatically. In fact, the semantics of each source can be discovered semi-automatically by data mining techniques. However, since the result is approximate, the presence of the administrator to validate this result is recommended. This stage is optional whether each source is already associated with an ontology.

Spatio-temporal ontologies [4,5] can be adapted to model spatio-temporal source in general, or trajectory ontologies [6–9] can be used for trajectory data in particular. These ontologies usually reuse an ontology of time and an ontology of space to represent the temporal and spatial dimension while the semantic one is modeled by a domain ontology. The OWL-Time ontology[1] [10] is a favorite candidate to represent temporal dimension. Regarding spatial dimension, one can reuse available models such as the GeoRSS model[2] or NeoGeo Geometry Ontology[3]. Transforming a spatial data model to spatial ontology is another possibility [6]. Ontologies proposed by query language (i.e., GeoSPARQL or stSPARQL) are also used to represent spatial data.

Ontologies Integration

This stage aims to establish semantic relations (i.e., equivalence, subsumption) between elements of the various ontologies. Numerous techniques have been proposed to discover the relations between terminologies and to take into account the differences between them. We can distinguish two main categories: ontology mapping and ontology fusion [11].

Data Integration

The final stage is based on constructed ontologies and their relations established in previous stages. It can be accomplished in two ways: either it builds mediators for virtual systems (on-demand mapping) or it populates data in a warehouse for materialized systems. This stage is detailed below.

Ontologies can be used as a conceptual schema over which queries are formulated. During the semantic data integration stage, ontologies are related to the databases via mappings that associate each ontological term with the underlying data. However, mapping development has received much less attention. Moreover, existing mappings are typically tailored to relate generic ontologies to a specific database schema. As a result, in contrast to ontologies, mappings usually cannot be reused across integration scenarios. Thus, each new integration scenario essentially requires the development of mappings from scratch [12].

Mappings can be used either for data materialization or on-demand mapping:

- **On-demand mapping:** In this approach, data remain located in databases; semantic queries must be rewritten into SQL ones at query evaluation step (if the source database is a relational model, which is not always the case).

---

[1] OWL-Time ontology:
  http://www.w3.org/2006/time.

[2] http://www.georss.org/.

[3] http://geovocab.org/geometry.

The approach is well suited in the context of very large datasets that would hardly support centralization due to resource limitations. Since no copy of the RDF data is made, it also guarantees up-to-date data.

– **Materialization:** Like warehouse approaches, data materialization is the transformation of the source databases into a RDF graph that is loaded into a triple store and accessed through a SPARQL[4] query engine. The whole process is often referred to as the Extract-Transform-Load (ETL) approach. The major advantage of the materialization is to facilitate further processing, analysis or reasoning on the RDF data. Since the RDF data are made available at once, third-party reasoning tools can be used to infer complex entailments. Furthermore, complex queries can be answered without compromising run-time performances since the reasoning has been performed at an earlier stage [13].

Several tools, such as D2RQ[5] [14], Datalift[6] [15], Morph-RDB[7] [16] or Ontop[8] [17] can be used to accomplish data translation. Figure 1 illustrates how to use D2RQ to translate heterogeneous data following the two approaches above.

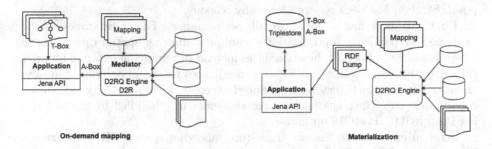

Fig. 1. Heterogeneous data translation with D2RQ.

The materialization approach is more appropriated for spatio-temporal data integration due to its best performances in querying and reasoning tasks. Furthermore, a geospatial triplestore can also be used to discover spatial relations that are represented by extensions of the querying language, which is not yet possible with current on-demand mapping-based system.

3   A Geospatial-Triplestore-Based Knowledge Base

A knowledge base comprises two components, the T-Box and the A-Box. The T-Box introduces the terminology, i.e., the vocabulary of an application domain,

---

[4] http://www.w3.org/TR/sparql11-query/.

[5] http://d2rq.org/.

[6] https://datalift.org/.

[7] http://mayor2.dia.fi.upm.es/oeg-upm/index.php/en/technologies/315-morph-rdb/.

[8] http://ontop.inf.unibz.it/.

while the A-Box contains assertions about named individuals regarding this vocabulary. In semantic data integration, the T-Box can be constructed by ontologies that model heterogeneous sources; and the A-Box is formed by the data translation process. When the knowledge base is defined, on the one hand, inferences can be applied to discover new relationships between translated data; on the other hand, additional experts knowledge can be injected for further investigation. The management of the knowledge base and especially the discovery of spatial relations can be realized by a geospatial triplestore.

3.1   Geospatial Triplestore Exposing a SPARQL Endpoint

Triplestores are DBMS for data modeled in RDF. Currently, several triplestores support storing and querying spatial data using GeoSPARQL or stSPARQL, extensions of SPARQL language. Concerning the use of the spatial dimension, and to our best knowledge, the best open-source triplestores for the spatial dimension support are Parliament[9] [18] and Strabon[10] [19]. Other triplestores support only a few type of geometries and geospatial functions [20]. In our proposal, Strabon has been selected for many reasons.

First, Strabon has a good overall performance. This advantage can be explained by particular optimization techniques allowing spatial operations to take advantage of PostGIS functionalities instead of relying on external libraries [21]: it pushes the evaluation of spatial predicates to the underlying spatially-enabled DBMS which has been enhanced recently with selectivity estimation capabilities [20]. Thus, spatial joins are also efficiently handled by the underlying PostgreSQL/PostGIS optimizer.

Strabon extends the Sesame triplestore, allowing spatial RDF data stored in the PostgreSQL DBMS enhanced with PostGIS. The triplestore works over the stRDF data model [19], a spatio-temporal extension of RDF in which the *strdf:WKT* and *strdf:GML* datatypes are introduced to represent serialized geometries using the two OGC standards: WKT and GML. The *strdf:geometry* which is the union of *strdf:WKT* and *strdf:GML* and the appropriate relationships are also introduced to represent the serialization of a geometry independently of the standard used for serialization.

The stSPARQL language used in Strabon has certain limitations, as it does not enable binary topological relations to be used as RDF properties. However, stSPARQL extends SPARQL 1.1 and overtakes GeoSPARQL by offering spatial aggregate functions and update commands of the triples. SQL functions can be utilized in SPARQL queries by defining a URI for each of them. Similarly, a boolean SPARQL extension function has been defined for each topological relation in three models: the OGC Simple Features, 9IM model, and RCC8 model. Thus, stSPARQL supports multiple families of topological relations and can express spatial selections and spatial joins. Another benefit of stSPARQL

---

[9] http://parliament.semwebcentral.org/.
[10] http://strabon.di.uoa.gr/.

is the support of federated queries and update statements that facilitate the updates of the knowledge base.

Another essential function of Strabon is the SPARQL endpoint. It is the interface that helps to access the content of Strabon by allowing users to formulate complex queries in the stSPARQL query language. The Web interface also provides an additional possibility to manage the knowledge base RDF data, for example, storing and updating RDF data.

3.2   Knowledge Base Enrichment

To exploit the knowledge base, additional information is required, such as the spatio-temporal relations between entities coming from heterogeneous sources or experts knowledge. This information can be used to enrich integrated data.

Spatio-Temporal Relations

Inferences are made based on the existing data and some additional information in the form of vocabulary, e.g., a set of rules. The Semantic Web Rule Language (SWRL[11]) is commonly chosen due to its available libraries, called *built-ins*, that provide several predicates, mostly for date-time and duration processing. Another approach is the application of SPARQL queries (SPARQL Construct or SPARQL Update) as a rule language. This approach is chosen as a reasoning mechanism for non-spatial data in our system.

In fact, at the current time, spatial relations cannot be inferred with simple rules. Several studies have introduced an SWRL *built-ins* to represent and infer spatial relationships between spatio-temporal objects but there are still limitations mainly with regard both to the system performance and reuse capabilities. Reasoning systems can be used to extract spatial relations from a knowledge base and reason over both these topological and directional relations through specialized software, such as JTS Topology Suite, but this approach impacts the system re-usability and data sharing. Therefore, in our project, the discovering of spatial relations is performed by the spatial functions of the geospatial triplestore.

Experts Knowledge

Experts knowledge can be added to improve the quality of integrated data. For example, subsumption or equivalence between knowledge base elements can be defined. These elements can also be linked to external sources, such as Linked Data[12], for additional information. Besides, experts knowledge can be used as semantic annotations for specific purposes.

---

[11] http://www.w3.org/Submission/SWRL/.
[12] http://linkeddata.org/.

4   Exploiting Heterogeneous Spatio-Temporal Data Through a Web Interface

After the implementation of a knowledge base tailored for spatio-temporal inference, we have proposed and developed a Web interface for exploiting heterogeneous spatio-temporal data.

First, the user connected to the Web server has to register the set of ontologies used to model the data sources. This process builds the T-Box of the knowledge base. Then, she has to load the mappings between ontologies and sources schemas to populate the A-Box. Afterwards, the knowledge base can eventually be enriched through spatio-temporal or business rules. Finally, the user can formulate semantic queries expressed in stSPARQL to exploit integrated data.

The above functionalities are provided by a framework consisting of three principal layers: the ETL, the management, and the application layer as shown in Fig. 2.

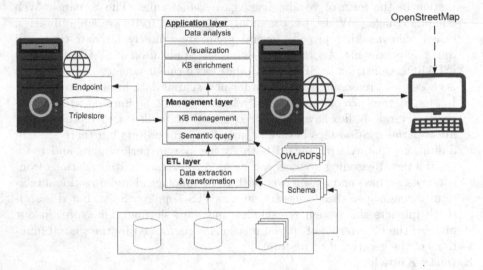

Fig. 2. A system architecture for exploiting spatio-temporal heterogeneous data.

ETL Layer

The layer manages mapping files that define how to connect to databases and how to match ontologies to databases schema. Being converted from relational data through the RDF conversion module, RDF triples can next be imported into the knowledge base. In this layer, the D2RQ framework[13] is used as a translation tool.

---

[13] D2RQ framework: http://d2rq.org/.

Management Layer

Two modules are developed. The *semantic query* module is used to process SPARQL semantic queries that are next sent the triplestore. The *KB management* module is used to expose the shared ontologies that mediate heterogeneous relational databases. Domain and spatio-temporal ontologies can be registered through this module. The essential tool used in this layer is the Jena framework[14].

Application Layer

It is composed of three modules: the *Visualization*, the *KB-enrichment* and the *data analysis* modules. The two latter are both referred to the management layer. In the *KB-enrichment* module, new statements can be inferred from spatio-temporal and business rules, while in the *data analysis*, an Apache Tomcat server is hosted to receive stSPARQL queries in the form of HTTP requests. The result is represented on a map by the OpenLayers library[15] with geographical data from OpenStreetMap[16]. Besides, statistical results may be visualized through dynamic diagrams thanks to Google chart tools[17]. The results can also be exported in CSV or JSON format for further analysis.

5   Application in Two Different Use Cases

At present, the system has been successfully applied to integrate and exploit spatio-temporal heterogeneous data of two projects.

GEMINAT (Environment and Landscape Geo-Knowledge)

This interdisciplinary project sets out to improve the reuse of environmental datasets collected inside the "Plaine & Val de Sèvre" workshop observatory since 1994. In this project, heterogeneous environmental data, such as crop rotation for 22 years of agricultural parcels (timestamped polygonal representation), and birds observations (timestamped point representation) are integrated through a spatio-temporal ontology. Once the knowledge base is built, experts knowledge may be added. For example, since there exists a lot of land-use, several methods of land use regrouping can be introduced according to the need for analysis. Used as semantic annotations, additional knowledge helps to improve the analysis quality.

The knowledge base is used to discover the relations between species observations and the land use of parcels. Experts can seek for animals preferences either by the type and form of land use. They can also verify the co-appearance of species according to the food chain or season. Figure 3 represents a view of the interface allowing for the co-analysis of raptors (the Gray Harrier in this case) abundance observations and crop rotation. In this

---

[14] Jena framework: http://jena.apache.org/.
[15] OpenLayers library: http://openlayers.org/.
[16] OpenStreetMap website: http://www.openstreetmap.org.
[17] https://developers.google.com/chart/.

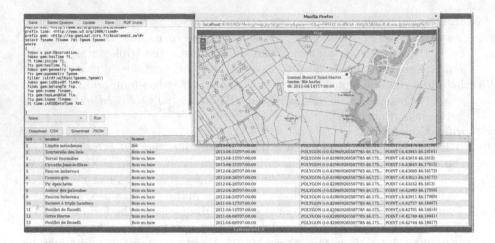

Fig. 3. Analyzing species observations with crop rotation data.

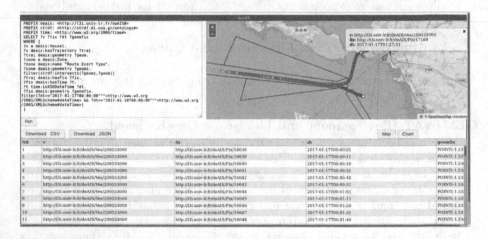

Fig. 4. Examine vessels arriving at La Rochelle ports on 11/27/2017.

analysis, the *inside* temporal relation and the *within* topological relation are used to link two data sources. For further details, see [22]. The system returns a set of triplets in less than 5 seconds on a two cores virtual server with 16 GB of RAM, reading more than 400 000 tuples describing the parcels, and 200 000 for the bird observation. Those performances allow for a high interactivity level.

déAIS (Detection of Faked AIS messages)

This project proposes a semantic approach for modeling, analyzing and detecting new maritime risks. The objective is to detect whether ship's AIS reports have been falsified (or spoofed) [23]. In the project, spatio-temporal heterogeneous maritime data, such as traffic data from AIS messages, coastal zones information or weather and tide information (work in progress) are integrated to annotate the behavior and activity of vessels. Additional information is linked to vessels trajectories by their spatio-temporal relations for a better understanding of the annotated behavior. Figure 4 represents a view of the system which helps to examine vessels arriving at La Rochelle ports. In this analysis, the *intersects* topological relation between vessels trajectories and zones ports are used. Again, good performances are recorded: less than 5 seconds for a set of 100 000 ships in the knowledge base.

6   Conclusion and Perspectives

A Web interface for exploiting spatio-temporal heterogeneous data is presented in this paper. The whole architecture is based on A-box and T-box materialization of domains ontologies mapped with original heterogeneous sources. This proposal insists on the spatio-temporal inference capabilities of the platform, that can't be achieved so easily nowadays due to performances cost of spatial reasoning. This justifies the choice of the materialization of data in the Strabon triplestore, and we show with the implementation of a Web server how the performances gain allows for an interactive analysis through a Web mapping client.

As a proof of concept, the system has been used to integrate and exploit heterogeneous data sources in two different projects. The client interface aims at more attractivity with an interactive map, charts, diagrams and a tabular presentation of requests. Users can also export the results. However, for the moment, queries are expressed in SPARQL, which is a language for computer scientists. To give more autonomy to the user, it would be now required to encapsulate the interface in tools allowing for a natural language like [24] or [25].

We believe that the system can be applied to others spatio-temporal data sources as long as they are modeled by ontologies. Future work concentrates on incorporating data mining techniques to facilitate the exploitation of integrated data.

References

1. Lenzerini, M.: Data integration: a theoretical perspective. In: Proceedings of the Twenty-First ACM SIGMOD-SIGACT-SIGART Symposium on Principles of Database Systems, PODS 2002, pp. 233–246. ACM, New York (2002)
2. Gray, A.J.G., Gray, N., Ounis, I.: Can RDB2RDF tools feasibily expose large science archives for data integration? In: Aroyo, L., Traverso, P., Ciravegna, F., Cimiano, P., Heath, T., Hyvönen, E., Mizoguchi, R., Oren, E., Sabou, M., Simperl, E. (eds.) ESWC 2009. LNCS, vol. 5554, pp. 491–505. Springer, Heidelberg (2009). https://doi.org/10.1007/978-3-642-02121-3_37
3. Cruz, I.F., Xiao, H.: The role of ontologies in data integration. J. Eng. Intell. Syst. **13**, 245–252 (2005)

4. Batsakis, S., Petrakis, E.G.M.: SOWL: spatio-temporal representation, reasoning and querying over the semantic web. In: Proceedings of the 6th International Conference on Semantic Systems, I-SEMANTICS 2010, pp. 15:1–15:9. ACM, New York (2010)
5. Harbelot, B., Arenas, H., Cruz, C.: Continuum: a spatiotemporal data model to represent and qualify filiation relationships. In: Proceedings of the 4th ACM SIGSPATIAL International Workshop on GeoStreaming, IWGS 2013, pp. 76–85. ACM, New York (2013)
6. Mefteh, W., Bouju, A., Malki, J.: Une approche ontologique pour la structuration de données spatio-temporelles de trajectoires. Application à l'étude des déplacements de mammifères marins. Revue Internationale de Géomatique **22**(1), 55–75 (2012)
7. Vandecasteele, A., Napoli, A.: An enhanced spatial reasoning ontology for maritime anomaly detection. In: 7th International Conference on System Of Systems Engineering - IEEE SOSE 2012, pp. 247–252, Genoa, Italy (2012)
8. Arenas, H., Harbelot, B., Cruz, C.: A semantic analysis of moving objects, using as a case study maritime voyages from eighteenth and nineteenth centuries. In: The Sixth International Conference on Advanced Geographic Information Systems, Applications, and Services, Barcelona, Spain, March 2014
9. Mohammadi, M.S., Isabelle, M., Thérèse, L., Christophe, F.: A semantic modeling of moving objects data to detect the remarkable behavior. In: AGILE 2017, Wageningen, Netherlands, May 2017. Wageningen University, Chair group GIS & Remote Sensing (WUR-GRS) (2017)
10. Hobbs, J.R., Pan, F.: An ontology of time for the semantic web. ACM Trans. Asian Lang. Inf. Process. **3**, 66–85 (2004)
11. de Bruijn, J., Ehrig, M., Feier, C., Martíns-Recuerda, F., Scharffe, F., Weiten, M.: Ontology Mediation, Merging, and Aligning, pp. 95–113. Wiley, Chichester (2006)
12. Pinkel, C., Binnig, C., Jiménez-Ruiz, E., May, W., Ritze, D., Skjæveland, M.G., Solimando, A., Kharlamov, E.: RODI: a benchmark for automatic mapping generation in relational-to-ontology data integration. In: Gandon, F., Sabou, M., Sack, H., d'Amato, C., Cudré-Mauroux, P., Zimmermann, A. (eds.) ESWC 2015. LNCS, vol. 9088, pp. 21–37. Springer, Cham (2015). https://doi.org/10.1007/978-3-319-18818-8_2
13. Michel, F., Montagnat, J., Zucker, C.F.: A survey of RDB to RDF translation approaches and tools. Research report, I3S, May 2014. ISRN I3S/RR 2013–04-FR, 24 p. (2014)
14. Bizer, C.: D2RQ - treating non-RDF databases as virtual RDF graphs. In: Proceedings of the 3rd International Semantic Web Conference (ISWC 2004) (2004)
15. Scharffe, F., Atemezing, G., Troncy, R., Gandon, F., Villata, S., Bucher, B., Hamdi, F., Bihanic, L., Képéklian, G., Cotton, F., Euzenat, J., Fan, Z., Vandenbussche, P.-Y., Vatant, B.: Enabling linked data publication with the Datalift platform. In: Proceedings of AAAI Workshop on Semantic Cities, Toronto, Canada, July 2012. scharffe2012a
16. Priyatna, F., Corcho, O., Sequeda, J.: Formalisation and experiences of R2RML-based SPARQL to SQL query translation using morph. In: Proceedings of the 23rd International Conference on World Wide Web, WWW 2014, pp. 479–490. ACM, New York (2014)
17. Calvanese, D., Cogrel, B., Komla-Ebri, S., Kontchakov, R., Lanti, D., Rezk, M., Rodriguez-Muro, M., Xiao, G.: Ontop: answering SPARQL queries over relational databases. Semant. Web **8**(3), 471–487 (2017)

18. Battle, R., Kolas, D.: Enabling the geospatial semantic web with parliament and GeoSPARQL. Semant. Web **3**(4), 355–370 (2012)
19. Kyzirakos, K., Karpathiotakis, M., Koubarakis, M.: Strabon: a semantic geospatial DBMS. In: Cudré-Mauroux, P., Heflin, J., Sirin, E., Tudorache, T., Euzenat, J., Hauswirth, M., Parreira, J.X., Hendler, J., Schreiber, G., Bernstein, A., Blomqvist, E. (eds.) ISWC 2012, Part I. LNCS, vol. 7649, pp. 295–311. Springer, Heidelberg (2012). https://doi.org/10.1007/978-3-642-35176-1_19
20. Garbis, G., Kyzirakos, K., Koubarakis, M.: Geographica: a benchmark for geospatial RDF stores (Long Version). In: Alani, H., Kagal, L., Fokoue, A., Groth, P., Biemann, C., Parreira, J.X., Aroyo, L., Noy, N., Welty, C., Janowicz, K. (eds.) ISWC 2013, Part II. LNCS, vol. 8219, pp. 343–359. Springer, Heidelberg (2013). https://doi.org/10.1007/978-3-642-41338-4_22
21. Patroumpas, K., Giannopoulos, G., Athanasiou, S.: Towards geospatial semantic data management: strengths, weaknesses, and challenges ahead. In: Proceedings of the 22Nd ACM SIGSPATIAL International Conference on Advances in Geographic Information Systems, SIGSPATIAL 2014, pp. 301–310. ACM, New York (2014)
22. Tran, B.-H., Bouju, A., Plumejeaud-Perreau, C., Bretagnolle, V.: Towards a semantic framework for exploiting heterogeneous environmental data. Int. J. Metadata Semant. Ontol. **11**(3), 191–205 (2016)
23. Ray, C., Iphar, C., Napoli, A., Gallen, R., Bouju, A.: DeAIS project: detection of AIS spoofing and resulting risks. In: MTS/IEEE OCEANS 2015, Gênes, Italy. IEEE, May 2015
24. Boumechaal, H., Allioua, S., Boufaïda, Z.: Conversion des requêtes en langage naturel vers NRQL. In: CIIA (2009)
25. Pradel, C., Haemmerlé, O., Hernandez, N.: Demo: Swip, a semantic web interface using patterns. In: The 12th International Semantic Web Conference (ISWC 2013), pp. 1–4, Sydney, Australia, October 2013

Increasing Maritime Situation Awareness via Trajectory Detection, Enrichment and Recognition of Events

G. A. Vouros[1], A. Vlachou[1], G. Santipantakis[1], C. Doulkeridis[1],
N. Pelekis[1], H. Georgiou[1], Y. Theodoridis[1], K. Patroumpas[1],
E. Alevizos[2], A. Artikis[1,2], G. Fuchs[3], M. Mock[3], G. Andrienko[3],
N. Andrienko[3], C. Claramunt[4(✉)], C. Ray[4], E. Camossi[5],
and A.-L. Jousselme[5]

[1] University of Piraeus, Piraeus, Greece
[2] NCSR 'D', IIT, Agia Paraskevi, Greece
[3] Fraunhofer Institute IAIS, Sankt Augustin, Germany
[4] Naval Academy Research Institute, Brest, France
christophe.claramunt@ecole-navale.fr
[5] CMRE, La Spezia, Italy

Abstract. The research presented in this paper aims to show the deployment and use of advanced technologies towards processing surveillance data for the detection of events, contributing to maritime situation awareness via trajectories' detection, synopses generation and semantic enrichment of trajectories. We first introduce the context of the maritime domain and then the main principles of the big data architecture developed so far within the European funded H2020 datAcron project. From the integration of large maritime trajectory datasets, to the generation of synopses and the detection of events, the main functions of the datAcron architecture are developed and discussed. The potential for detection and forecasting of complex events at sea is illustrated by preliminary experimental results.

Keywords: Big Spatio-temporal data · Moving objects · Trajectory detection
Data integration · Events recognition/forecasting

1 Introduction

The maritime sector is growing and currently employs around 5.4 million people in Europe, with a value estimated in 500 billion Euros at year for Blue Growth activities. Maritime traffic is constantly increasing, likewise the exploitation of sea resources. To improve safety of navigation and sustain the development of the so-called Blue economy, maritime surveillance systems should support authorities in processing larger amount of heterogeneous data and monitoring efficiently larger areas. Indeed, existing systems are not able to fully support Maritime Situation Awareness (MSA), which requires the correlated use of large, heterogeneous and uncertain data sources. The amount of data to be correlated, as well as their variety in formats and characteristics, is

M. R. Luaces and F. Karimipour (Eds.): W2GIS 2018, LNCS 10819, pp. 130–140, 2018.
https://doi.org/10.1007/978-3-319-90053-7_13

unsustainable for traditional systems, which are now required to face all the challenges of Big Data at once.

The datAcron project considered and defined maritime scenarios [2] that address operational concerns regarding fishing activities, highlighting the need for continuous (real-time) tracking of fishing vessels and surrounding traffic, as well as contextually enhanced offline data analytics. The secure fishing scenarios are designed to demonstrate our ability to detect and foresee situation indicators regarding collisions between ships and vessels in distress optimizing rescuing operations. In addition to these, we are particularly interested on maritime sustainable development scenarios, where we aim at monitoring the impact of fishing activities, including the illegal ones. In particular, the protection of areas from fishing scenario tackles Illegal Unreported Unregulated (IUU) fishing, which is a global threat to the preservation of maritime ecosystems and could potentially undermine the sustainable development in large areas of the world that depend on maritime resources. In this scenario, we aim to support authorities dealing with real-time monitoring of protected areas and areas where fishing is restricted, by predicting and detecting vessels entering the surveyed areas. The user needs to forecast whether and when a vessel enters, exits, sails or spends time in such areas.

This paper aims to show the deployment and use of advanced technologies developed by the datAcron project (www.datacron-project.eu) towards events' recognition in the protection of areas from fishing scenario. The objective is to demonstrate the part of the overall datAcron architecture that processes surveillance data for the detection of events, contributing to maritime situation awareness via trajectories' detection, synopses generation and semantic enrichment of trajectories. The rest of the paper is organized as follows. First, Sect. 2 describes the overall datAcron big data architecture. Next, the trajectory enrichment and detection, as well as events' forecasting: Technologies to be demonstrated are described in Sects. 3 and 4. Finally Sect. 5 draws the conclusions and outline further work.

2   A Big Data Architecture for Time Critical Mobility Forecasting

Time critical mobility operations in the maritime domain require integrating data that stems from a wide variety of diverse data sources, both archival (data-at-rest) and online (data-in-motion), which is also voluminous and produced at high rates. During data acquisition, various tasks need to be performed, including data cleaning, compression, transformation to a common representation model, data integration and interlinking. Besides real-time operations that must be supported with minimum latency requirements (i.e., in real-time), there exists a need for offline analysis to extract useful knowledge.

The datAcron system architecture, depicted in Fig. 1, can be considered as a Big Data architecture for processing both real-time and archival data. While it bears similarities with the Lambda architecture [9], since it encompasses both a real-time and a batch processing layer, these layers exist for different purposes (e.g., online trajectory/events forecasting vs offline trajectory clustering and visual analytics over archival data).

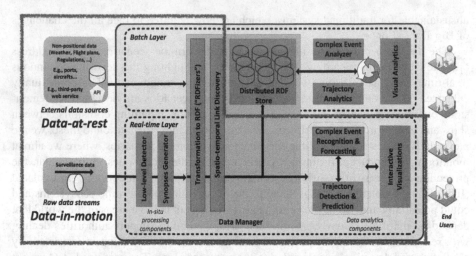

Fig. 1. The overall datAcron system architecture and the demonstrated components indicated by the cover (red) polygon. (Color figure online)

The real-time layer, which is the subject of this paper, involves feeding into the system streaming surveillance data describing the positions of moving objects, collected from terrestrial and satellite receivers. While consuming this data, statistics (min/max/avg) are computed over properties, such as speed and acceleration, in an online fashion; online data cleaning of erroneous data, and trajectory compression, are performed. Then, compressed trajectories (i.e., trajectory synopses) generated are transformed to RDF, according to the datAcron ontology [12], thereby facilitating the computation of links with relevant data originating from other sources. To this end, spatio-temporal link discovery is performed that discovers relations between surveillance data and archival data (e.g., weather, contextual data), resulting in enriched trajectories. Further online analysis of enriched trajectories is performed, aiming at: (a) deriving predictions of the future location of a moving object, and (b) complex event recognition and forecasting. Finally, real-time visualizations support human interaction with the datAcron system.

In the batch layer, both the enriched trajectories as well as data from other sources that have been transformed in RDF are collected for persistent storage, in order to support offline data analytics. Due to the immense data volume, parallel data processing is performed over RDF data stored in a distributed way. On top of the distributed RDF store, high-level data analysis tasks run, in order to perform trajectory analysis as well as building models for complex event forecasting using machine learning techniques. Visual analytics provide the ability to discover hidden knowledge and patterns, by means of interaction with a domain expert or a data analyst.

The big data technologies employed for the implementation of the architecture include a blend of state-of-the-art solutions that are used in production environments successfully: Stream processing components have been developed in Apache Flink, harnessing the scalability and low latency offered. For batch processing and analysis, we have selected Apache Spark which is the most popular batch processing framework to-date, achieving scalability, high performance, and exploiting in-memory processing. The stream-based communication between components is achieved by means of Apache Kafka.

This article concerns the streaming layer and describes the processing stages performed from congesting surveillance data to visualizations of trajectories detected and of events recognized and forecasted. However, it is separated in two parts: trajectory detection and complex events recognition and forecasting.

3 Trajectory Detection

The section describes the functionalities of components involved in the data integration and enrichment part of the overall framework: (a) The in-situ processing component for cleansing and enrichment of surveillance data with derived information (e.g. average speed), as well as for the detection of low-level events; (b) trajectories' synopses generation from raw streaming surveillance data via the detection and tagging of critical points, and (c) RDF generation of trajectory synopses and their linking to other data sources for the provision of enriched trajectories.

3.1   In-Situ Processing and Low-Level Events Detection

The low-level event detection component is aiming at enriching the raw-data generated by the moving entities with basic derived attributes that serve as input for higher-level processing. A major consideration in this low level is to achieve enrichment with low-latency, preferably by so-called "in-situ" processing. In-situ processing refers in general to the case of processing streaming data as "downwards" in-stream as possible. Processing streaming data close to data source provides a number of inherent advantages, such as decreased communication delays, savings in communication, and reduced overhead in sub-sequent evaluation steps. The low-level events refer to two basic datAcron tasks to be performed in real-time on the trajectories: generating metadata on incoming raw data for detection of erroneous data and ensuring data quality, and enriching the data stream with contextual information for further analysis. For supporting the data quality assessment, attributes of min/max, median/average of properties (e.g. speed, acceleration, etc.) are generated on a per trajectory basis. In addition to that, raw position data are enriched with low-level events of entering or leaving of moving entities from one area to another one, by processing the real-time stream of moving entity positions.

3.2   Synopses Generation

Detecting important mobility events along trajectories has to be carried out in a timely fashion against the streaming positional updates received from a large number of vessels. Instead of retaining every incoming position for each object, we have implemented a Synopses Generator module that drops any predictable positions along trajectory segments of "normal" motion characteristics, since most vessels usually follow almost straight, predictable routes at open sea. Indeed, a large amount of raw positional updates may be suppressed, while only retaining locations that signify changes in actual motion patterns [15]. We opt to avoid costly trajectory simplification algorithms like [6, 7] operating in batch fashion, online techniques employing sliding windows [8], or safe area bounds for choosing samples [7], as well as more complex, error-bounded methods. Instead, emanating from the novel trajectory summarization framework introduced in [11] for online maritime surveillance, but significantly enhanced with additional noise filters, the Synopses Generator applies single-pass heuristics for achieving succinct, lightweight representation of trajectories. We prescribe that each trajectory can be approximately reconstructed from judiciously chosen critical points of the following types:

- Stop: It indicates that an object remains stationary (i.e., not moving) by checking whether its instantaneous speed is lower than a threshold (e.g., 0.5 knots) over a period of time.
- Slow motion: It signifies that an object consistently moves at low speed (e.g., <5 knots) over a period of time.
- Change in Heading: Once there is an angle difference in heading of more than a given threshold (e.g., >5°) with respect to the mean velocity vector (computed over the most recent course of a given object), its current location should be emitted as critical.
- Speed change: Such critical points are issued once the rate of change for speed exceeds a given threshold (e.g., >25%) with respect to its mean speed over a recent time interval.
- Communication gaps: These occur when an object has not emitted a message over a time period, e.g., the past 10 min.

Critical points can be emitted at operational latency (i.e., within milliseconds) and high throughput. Hence, this derived stream of trajectory synopses can keep in pace with the incoming raw streaming data. This module can also achieve dramatic compression over the raw streaming data with tolerable error in the resulting approximation. At lower or moderate input arrival rates, data reduction is quite large (around 80% with respect to the input data volumes), but in few cases of very frequent position reports, compression ratio can even reach 99% without harming the quality of the derived trajectory synopses.

3.3   RDF Generation and Data Integration

The next step of the data processing workflow is to convert the synopses generated to RDF and integrate them to archival data into a knowledge graph. Since several different sources are blended into our domain, we designed and implemented a generic RDF generation framework. Triples generated from the RDF generators are directed to a group of Link Discovery components.

The proposed method stands on two main components: (a) the data connector, and (b) the triple generator. The data connector is responsible to connect to a data source and accept the data provided. It applies naive data cleaning, computes and converts values, applies simple filters, and generates values from the incoming entries, e.g. extracting the Well-Known-Text representation of a given geometry in a Shapefile. The output of these connectors is directed to instances of the triple generator component.

The triple generator is responsible to convert all the data coming through the data connector, into meaningful triples w.r.t. the datAcron ontology [12]. This component depends on the use of graph templates and variable vectors. The variables vector enables transparent reference to variables and use of their values. The graph template on the other hand, uses these variables into triple patterns, i.e. triples where any of the subject or object can be either a variable or a function with variable arguments.

In contrast to other RDF generators, the proposed method needs no further knowledge of a specific vocabulary (e.g. compared to RML [3]), and it can be used by anyone who can write simple SPARQL queries. Furthermore, it requires no underlying SPARQL engine, and it inherently supports parallelization and streaming data sources (e.g. compared to SPARQL-Generate [5] and GeoTriples [4]). In addition to these, the variables vector enables the RDF generation method to establish mappings to data "to-be-generated", and they are not explicitly available in the source (e.g., the MBR or the WKT of a geometry).

3.4 Link Discovery

The output of the RDF generators is further exploited for the detection of associations between entities, or the enrichment of the generated RDF graph with additional information from any of the sources available.

The link discovery component detects spatio-temporal and proximity relations such as "within" and "nearby" relations between stationary and/or moving entities. It is noteworthy that there is not much work on the challenging topic of spatio-temporal link discovery, nor on link discovery over streaming datasets. State of the art approaches such as [10, 13, 14] focus on spatial relations in static archival datasets only. In particular RADON [13] employs optimizations that can be only applied if the datasets are a-priori accessible as a whole, which cannot be assumed for streaming datasets. Our work addresses explicitly proximity and spatio-temporal relations in both archival and streaming data sources.

The implemented component continuously applies SPARQL queries on each RDF graph fragment produced by an RDF generator, to filter only those triples relevant to the computation of a relation. It applies a blocking method to organize entities (either being moving or stationary entities), and a refinement function to evaluate pairs of entities in any block.

Aiming to discover spatio-temporal relations among entities, methods use an equi-grid which organizes entities by space partitioning. The temporal dimension is not partitioned: given a temporal distance threshold, we can safely clean up data that are out of temporal scope, i.e. entities that will never satisfy the temporal constraints of the relations. To effectively prune candidate pairs of entities, the proposed method computes the complement of the union of those spatial areas that correspond to entities in a cell and intersect with the cell's area: This cell area is called the *mask* of cell.

Thus, for each new entity we identify the enclosing cell, and then we evaluate that entity against the spatial mask of the cell. If it is found to be in the mask, we do not need to further evaluate any candidate pair with entities in that cell. In addition to masks, the link discovery component uses a book-keeping process for cleaning the grid, towards identifying proximity relations among entities when dealing with streamed data.

4   Complex Events Recognition and Forecasting

This section shows the on-line recognition and forecasting of events which are visualized together with the visualization of enriched stream of trajectories. This involves the Complex Events' Recognition and Forecasting module consuming the enriched stream of trajectories' synopses and streaming out events. Besides the critical points generated by the synopses generator, this module also consumes, in the form of events, extra information provided by the link discovery component, especially the spatial relations between vessels and areas.

Given the enriched stream of synoptic trajectories (i.e. streams of trajectory critical points linked with low level events, weather features and contextual information) and a set of patterns defining relations between low-level events, operational constraints and contextual information, we need to detect, in a timely manner, when patterns' relations (involving temporal and spatial aspects) are satisfied. Whenever this happens, a high-level (complex) event has been detected. In addition, we need to forecast the occurrence of complex events.

5   Event Detection and Forecasting

As a first step, event patterns in the form of regular expressions are converted to deterministic finite automata (DFA). A detection occurs every time the DFA reaches one of its final states. As an example, Fig. 2a depicts the DFA constructed for the simple sequential expression R = acc, where events that may be encountered are $\Sigma = \{a, b, c\}$. For the task of forecasting, a probabilistic model need to be built for (the behavior of) the DFA. We achieved this by converting the DFA to a Markov chain. Assuming the input stream provides Independent and Identically Distributed (IID) low-level events, then it can be shown that we can directly map the states of the DFA to states of a Markov chain as well as the transitions of the DFA to transitions of the Markov chain. The probability of each transition would then be equal to the occurrence probability of the event that triggers the corresponding transition of the DFA. However, if we relax the assumption of IID events, then a more complex transformation is required, in which case the transition probabilities equal the conditional probabilities of the events. Figure 2b shows the Markov chain derived from the DFA illustrated in Fig. 2a, assuming that the input events are generated by a 1st-order

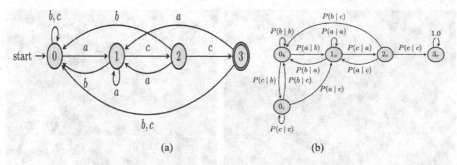

Fig. 2. (a) DFA and (b) Corresponding Markov Chain.

Markov process (refer to [1] for details). We call such a derived Markov chain a Pattern Markov Chain (PMC).

Once we have obtained the PMC corresponding to an initial pattern, we can compute certain distributions that are useful for forecasting. At each time point the DFA and the PMC will be in a certain state, the question we need to answer is the following: how probable is it that the DFA will reach its final state (and therefore a complex event will be detected) in k time steps from now (waiting-time distribution)?

Forecasts are provided in the form of time intervals, like I = (start, end). Such a forecast specifies that the DFA is expected to reach a final state in the future between *start* and *end* with a probability of at least a given constant threshold θ (provided by end-user). These intervals are produced by a single-pass algorithm that scans a waiting-time distribution and finds the smallest (in terms of length) interval that exceeds this threshold. This method has been implemented in the Scala programming language in a system called Wayeb.

5.1 Use Cases

Several maritime scenarios and related events have been defined in [2] and event patterns have been formalized and implemented. We evaluated event detection and prediction using real vessel data obtained through the Automatic Identification System (AIS). The dataset includes approximately 18 million of AIS positions transmitted by about 5,000 vessels sailing in the Atlantic Ocean around the port of Brest, France, between October 2015 and March 2016. Moreover, several navigation features such areas of interest, coastlines, ports locations... have been considered.

Amongst experiments realized, Fig. 3 illustrates events from one pattern applied to a single vessel and one area of interest. The aim is to predict when the vessel is expected to enter the area. This "within" event is of crucial importance for the early detection and prevention of possible collisions: Maritime experts need to know whether a cargo vessel is heading towards a fishing area, since this indicates a possibility of collision.

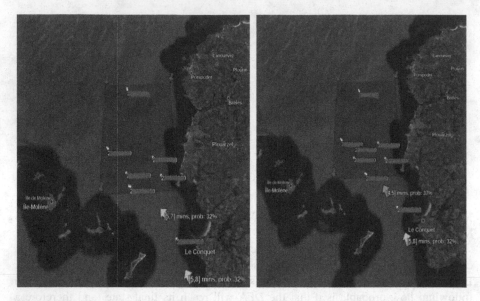

Fig. 3. Visualization of events forecasting [Google earth V 7.1. Mer d'Iroise, France, 48°23'08.76"N, 4°51'37.54"W. SIO, NOAA, U.S. Navy, NGA, GEBCO. DigitalGlobe 2015. http://www.earth.google.com, September, 2017]. (Color figure online)

We implemented this functionality by creating a pattern with a single event, namely the "within" event of the link discovery component. We subsequently set a high order for the derived PMC and we let this PMC learn by itself which sequences of events have a high probability of leading to a "within" event. For instance, a sequence of "close", "close", "very close", "very close" events (w.r.t. quantitative thresholds defined with experts), have a high probability of leading to a "within" relation (the vessel steadily approaches the area).

Figure 3 shows an area (red rectangle) and vessels (enlarged arrows) along two similar routes. The arrow informs us (left Figure) that the vessels are expected to enter the area in 5 to 7 min with a probability at least 32% (resp. 5 to 8 min, 32%). The other arrow in between has no such interval because it is on a route that does not cross the area (the vessel's identities have been erased for privacy reasons). Results showed that, the closer the vessel is to the area, the higher the precision becomes and the smaller the forecast intervals (Fig. 3, left). Additionally, when the vessel follows a route that does not normally cross the area, the forecasting module refrains from producing intervals, indicating that it has learnt that this route doesn't involve any "within" events.

The real-time layer of the datAcron architecture described in Fig. 1 includes a visualization interface supporting human interaction (on-going work). The aim of the interface is to provide visualization of low-level event (Sect. 3.1), critical points (Sect. 3.2) and complex events (Sect. 4). The visualization of events has been designed as a web-based interface showing the ships' tracks, real and predicted events. Figure 4 shows this interface and a few detected critical points.

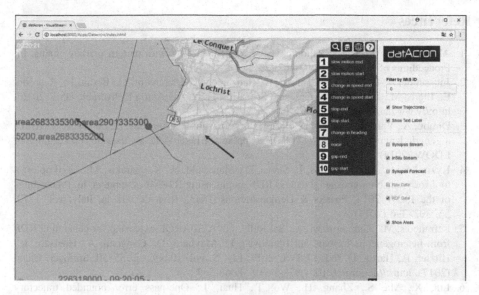

Fig. 4. Visualization dashboard for maritime events.

6 Conclusion

The research presented in this paper presents some preliminary results of the datAcron project whose objective is to advance the management and integrated exploitation of voluminous and heterogeneous data-at-rest (archival data) and data-in-motion (streaming data) sources, so as to significantly advance the capacities of systems to promote safety and effectiveness of critical operations for large numbers of moving entities in large maritime areas.

We introduce our current progress and achievements towards the real-time processing and analysis of big data for improving the predictability of trajectories and events regarding moving entities in maritime domain. There are still many challenges ahead to be addressed such as discovery of a interactions among moving ships in a timely manner, efficient query answering of very large knowledge graphs for online and offline analytics tasks, cross-streaming synopses generation at the data integration level, long-term online full trajectory predictions and improvements in forecasting complex events together with learning/refining their patterns by exploiting examples are amongst some of major challenges ahead we still plan to address.

Acknowledgments. This work was supported by project datACRON, which has received funding from the European Union's Horizon 2020 research and innovation program under grant agreement No 687591.

References

1. Alevizos, E., Artikis, A., Paliouras, G.: Event forecasting with pattern markov chains. In: Proceedings of DEBS, pp. 146–157 (2017)
2. Jousselme, A.-L., Ray, C., Camossi, E., Hadzagic, M., Claramunt, C., Bryan, K., Reardon, E., Ilteris, M.: Maritime Use Case and Scenarios, H2020 datAcron Deliverable D5.1 (2016). http://www.datacron-project.eu/
3. Dimou, A., Vander Sande, M., Colpaert, P., Verborgh, R., Mannens, E., de Walle, R.: RML: a generic language for integrated RDF mappings of heterogeneous data. In: Proceedings of LDOW (2014)
4. Kyzirakos, K., Vlachopoulos, I., Savva, D., Manegold, S., Koubarakis, M.: GeoTriples: a tool for publishing geospatial data as RDF graphs using R2RML mappings. In: Proceedings of the ISCW 2014, Posters & Demonstrations Track, Riva del Garda, Italy, vol. 1272, pp. 393–396 (2014)
5. Lefrançois, M., Zimmermann, A., Bakerally, N.: A SPARQL extension for generating RDF from heterogeneous formats. In: Blomqvist, E., Maynard, D., Gangemi, A., Hoekstra, R., Hitzler, P., Hartig, O. (eds.) ESWC 2017. LNCS, vol. 10249, pp. 35–50. Springer, Cham (2017). https://doi.org/10.1007/978-3-319-58068-5_3
6. Lin, X., Ma, S., Zhang, H., Wo, T., Huai, J.: One-pass error bounded trajectory simplification. PVLDB **10**(7), 841–852 (2017)
7. Liu, J., Zhao, K., Sommer, P., Shang, S., Kusy, B., Jurdak, R.: Bounded quadrant system: error-bounded trajectory compression on the go. In: Proceedings of ICDE, pp. 987–998 (2015)
8. Long, C., Wong, R.C.-W., Jagadish, H.V.: Trajectory simplification: on minimizing the direction-based error. PVLDB **8**(1), 49–60 (2014)
9. Marz, N., Warren, J.: Big Data - Principles and Best Practices of Scalable Real-Time Data Systems. Manning Publications, Greenwich (2015)
10. Ngonga Ngomo, A.-C.: ORCHID – reduction-ratio-optimal computation of geo-spatial distances for link discovery. In: Alani, H., et al. (eds.) ISWC 2013. LNCS, vol. 8218, pp. 395–410. Springer, Heidelberg (2013). https://doi.org/10.1007/978-3-642-41335-3_25
11. Patroumpas, K., Alevizos, E., Artikis, A., Vodas, M., Pelekis, N., Theodoridis, Y.: Online event recognition from moving vessel trajectories. GeoInformatica **21**(2), 389–427 (2017)
12. Santipantakis, G., Vouros, G., Doulkeridis, C., Vlachou, A., Andrienko, G., Andrienko, N., Fuchs, G., Garcia, J.M.C., Martinez, M.G.: Specification of semantic trajectories supporting data transformations for analytics: the datAcron ontology. In: Proceedings of SEMANTICS, pp. 17–24 (2017)
13. Sherif, M.A., Dreßler, K., Smeros, P., Ngomo, A.N.: Radon - rapid discovery of topological relations. In: Proceedings of AAAI 2017, pp. 175–181 (2017)
14. Smeros, P., Koubarakis, M.: Discovering spatial and temporal links among RDF data. In: Proceedings of LDOW (2016)
15. Bertrand, F., Bouju, A., Claramunt, C., Devogele, T., Ray, C.: Web architecture for monitoring and visualizing mobile objects in maritime contexts. In: Ware, J.M., Taylor, G.E. (eds.) W2GIS 2007. LNCS, vol. 4857, pp. 94–105. Springer, Heidelberg (2007). https://doi.org/10.1007/978-3-540-76925-5_7

From What and When Happen, to Why Happen in Air Pollution Using Open Big Data

Roberto Zagal-Flores[1], Miguel Felix Mata[1] ⓘ,
and Christophe Claramunt[2(✉)]

[1] Instituto Politécnico Nacional, UPIITA-IPN, Ciudad de México, Mexico
{rzagalf,mmatar}@ipn.mx
[2] Naval Academy Research Institute, Brest Naval, 29240 Brest, France
claramunt@ecole-navale.fr

Abstract. The air pollution phenomenon has been often studied from an environmental dimension but not from a spatial big data approach and considering social perception analysis. In order to understand such complex phenomenon a multidimensional analysis of heterogeneous environmental data might provide new insights. Notably, the Mexico government has released open data on air quality that contains the historical behavior of air pollution in Mexico City, while social networks data provides rich descriptions regarding regional social problems. In order to take into account the respective contributions of these two data sources from a spatial-temporal perspective, we introduce a multidimensional approach whose objective will be to integrate these heterogeneous data sources in an unified framework. While human perception often embedded in social media is naturally subjective, public data is rather objective and reliable, while they are described at different levels of temporal granularity and scale. Therefore, the search for a sound integration of these data sources is surely a non-straightforward issue. The research presented in this paper introduces a modelling and data mining approach to search for spatial-temporal patterns that can describe not only what happens, but also why such phenomenon happens. The whole framework is applied to the study of air pollution in Mexico City. The idea is to connect unstructured data (social data) and structured spatial data (open data) through the reconciliation of spatial-temporal correspondences between them to discover new geographic knowledge on Air Pollution phenomenon in Mexico City.

Keywords: Geographic knowledge discovery · Social perception
Spatial data mining

1 Introduction

The recent adoption of open data paradigms and the evolution of Big Data technologies have allowed analyzing large open data sources produced by many governments and organizations [1]. In particular, air pollution is a geographical-human generated phenomenon that generates many interactions between the geographical environment and social structures producing reliable but subjective data. Despite the intrinsic subjective component of social media data, there are a worthwhile range of geographical knowledge

© Springer International Publishing AG, part of Springer Nature 2018
M. R. Luaces and F. Karimipour (Eds.): W2GIS 2018, LNCS 10819, pp. 141–154, 2018.
https://doi.org/10.1007/978-3-319-90053-7_14

and social behavior patterns hidden in this data [2]. Nowadays, very large amounts of objective and subjective data are overall generated by sensors, governments, social media and the Web, this representing a very valuable source of information [3].

The Mexico City government shares air quality data related to carbon dioxide, fine particles and other air pollutants over a spatial-temporal scale of 16 boroughs and over a period of 30 years. Although this open data is reliable, it only represents the expert dimension of the environmental phenomenon [1]. In contrast, social networks become relevant as a means of diffusion and sharing of citizen views (i.e., the social dimension). At large, human perceptions are here generated when a large number of users express impeachments, complaints, opinions and locations related to air pollution topics [7, 8]. Indeed, human perception and opinion data is subjective and not always reliable and precise. Our assumption is that such data must be integrated with conventional data sources in order to provide a better understanding of an air pollution phenomenon from space-temporal and multidimensional approaches. The challenge relies in the underlying diversity of spatial-temporal scales, and the integration of related data sources like traffic, social protest and respiratory diseases [6], this clearly giving to the whole data integration process a route towards a big data science application [4, 5].

This paper presents the preliminary results of a geographic knowledge discovery methodology that uses social perception analysis, spatial-temporal approach and spatial big data paradigms. The idea is to connect unstructured data (social data) and structured spatial data (open data) through the reconciliation of spatial-temporal correspondences between data sources in order to discover new geographic knowledge on Air Pollution phenomenon in Mexico City. The main contribution relies on the crossing of different datasets to discover some geographic and temporal patterns as reflected by social media, institutional data sources and the final geographic social perception. This paper extends our preliminary works oriented to the discovery of geographical knowledge and patterns [11–14].

The remainder of the paper is organized as follows. Section 2 briefly introduces a survey of related works. Section 3 describes the methodology of our approach. Section 4 introduces the preliminary results while Sect. 5 concludes the paper and outlines further work.

2 Related Work

The study of related work is organized in three topics: (1) Geographic Knowledge Discovery (GKD) methodologies applied to air pollution and social phenomena, (2) Big Data architectures based on a multidimensional approach, (3) Spatio temporal models oriented to the analysis of data cubes. This should provide a sort of background to our work oriented to the integration of these research topics from a multidimensional approach that allows the integration of subjective and objective data.

GKD Methodologies: The work developed in [23] introduces a cloud-based knowl-edge discovery system that infers real-time and fine-grained air quality information throughout historical and real-time air quality data and others city data sources (e.g., meteorology, traffic flows, human mobility, road networks, points of interest). This system was evaluated using real data from 9 cities in China. However, the authors didn't consider social perception as dimension to discover social consequences and impacts related to the air pollution phenomenon. In [8] a methodology is oriented to the detection of human perception, it analyses the exposure of human opinions to the news, citizen-related information through social media, the goal being to discover influence exposure to perspectives that cut across ideological lines of the voters. Overall it appears form that study that geography surely plays a role in facilitating connections among voters with similar ideologies. The work in [18] analyzes the social perception and emotion-based distribution of tweets at a large country scale. The approach extracts and categorizes tweets based on semantic orientations of terms in a dictionary, and explores their spatial and temporal distribution.

Big Data Applications and Multidimensional Approach: The work in [25] presents a spatial model that investigates the impacts of both air pollution and its spatial spillover effect on public health in 116 cities of China. The authors use survey data of lung cancer mortality and respiratory diseases mortality from 161 sites of mortality cen-soring spots in China to measure public health and also integrate statistical data of industrial emissions of sulfur dioxide and soot from the corresponding cities to measure air pollution. This dataset can be regarded as a big data sample in the fields of envi-ronmental and health economics. The empirical results provide evidence for the adverse impacts of air pollution and its spatial spillover effect on public health. In [10] a Big Data architecture integrates non-traditional information sources and data analysis methods in order to provide forecasting social and economic behaviors, trends and changes. The authors do not use clustering in classified texts to discover terms asso-ciated to topics social perception.

The predictive model developed in [24] is a system that applies data-driven models to predict fine-grained air quality over a period of 48 h. This system connects air quality and meteorological data, and weather forecasts from over 3,000 sources in China. A series of spatial big data architectures have been also applied to social data phenomena, in [9] a review of related work and research challenges and opportunities in spatial big data are explored. Several case studies are introduced to show the importance and benefits of geospatial big data, including transportation planning, socio-economical trends, urban planning, and health care. However, the authors didn't report a big data architecture applied to the air quality domain.

Spatio-Temporal Models: Spatio-temporal analysis applied to Air Quality cubes are explored in [16], where the authors propose a multidimensional exploration and analysis methodology applied to textual and spatial representations of air pollution data in Italian regions. Data is organized according to different axes (i.e., space, time, pollution) and a hierarchical approach and facts that represent the subject of the analysis. The objective is to allow a decision maker to analyze in space and time the average value of the pollution by year and pollutants. The main idea developed in [17] is the use of data facts in the analysis and simulations using different dimensions through Spatial data cubes using a SOLAP approach (Spatial On-Line Analytical Processing). Although spatiotemporal and GKD (Geographical Knowledge Discovery) frameworks have been well studied in [6, 15], new approaches are still needed to reconcile structured and unstructured data. In fact, a unified view of such heterogeneous datasets requires novel mechanisms to manage exploration at different degrees of spatio-temporal scale and granularity.

This brief related work study considers methodologies and frameworks essentially or potentially applied to air pollution analysis using GKD and BigData paradigms, the concept of data dimensions and spatio-temporal analysis. The main assumption of our approach is to consider additional data sources related to air pollution. However, most of current related works presented above don't consider different data dimensions as well as how such phenomenon is reflected by human social perception. Instead we believe that such complementary social data provide a rich a source of knowledge that should allow to iteratively analyze such phenomenon.

3 Methodology: Tripartite Components in Spatial Big Data

This section introduces a spatial big data methodology based on a tripartite component: space, time and the social dimension. It combines a pipeline machine learning architecture (supervised and unsupervised methods) and a spatio-temporal analysis approach. Social data was recollected from verified user publications, Facebook and Twitter communities, and that contains opinions, complaints, and impeachments; the idea is to generate a social perception intuitive map. Air quality and road incidents (e.g., social protests, car accidents, etc.) open data sources were also considered. The main goal is to discover potential relations between social perception and open data. Therefore one objective of our approach will be to potentially answer this kind of question: What are the most important air quality topics according to the people? Does perception change according to the geographical location and time? Does social perception correspond to real events according to open data available? How does the air pollution phenomenon behave according to spatial-temporal properties of the open data?

Our approach adapts paradigms derived from Geographic Knowledge Discovery and Big Data approaches [6, 9]. The methodology uses four iterative phases: (1) Detection of geographical social perception using supervised machine learning techniques; (2) spatio-temporal mediation that applies an equalizer concept as a mediator; (3) multidimensional analysis based on an ontology that favors a cross-comparison of the incoming data; and (4) data visualization component that enables us to explore the unified data. The methodology is iterative because the discovered results can be used in a second discovery iteration. These different phases are illustrated in Fig. 1.

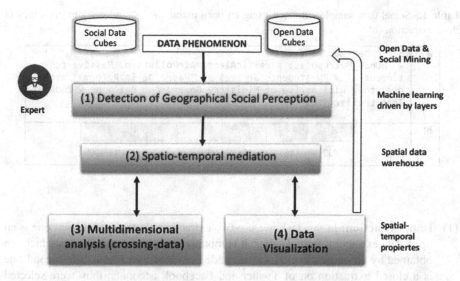

Fig. 1. Geographical knowledge discovery methodology

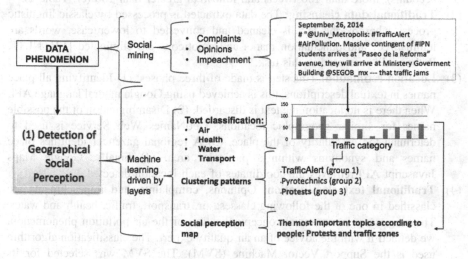

Fig. 2. An overview of the social perception detection.

3.1 Detection of Geographical Social Perception

This phase consists of four tasks: (1) data extraction definition; (2) traditional data cleansing; (3) location estimation; (3) text classification; and (4) text clustering. This phase was experimented and developed in our previous work [12, 13]. Figure 2 shows an overview of this phase.

Table 1. Social data sample extracted using an open extraction (a) and account properties in close extraction (b)

a)	September 26, 2014 # "@Univ_Metropolis: #TrafficAlert #AirPollution. Massive contingent of #IPN students arrives at "Paseo de la Reforma" avenue; they will arrive at Ministery Government Building @SEGOB_mx That traffic jams					
b)		Followers	Follows	Retweets	Favorites	Verified	User Since
		237	861	174	0	Yes	24/11/2009

(1) **Data extraction:** In this task, we use two extraction strategies. The first one is an open data extraction from Twitter; it combines key terms and hashtags which are obtained by an empirical exploration guide provided by experts. The second one is a closed extraction on of Twitter and Facebook accounts; they were selected considering users active since at least one year, verified account (e.g., government accounts), more than 100 tweets and followers greater than 100 (cf. Table 1).

(2) **Traditional data cleansing:** The data extracted is processed by classic linguistic process whereby the text is cleaned and converted to lower case, words are lemmatized, and punctuation marks and emoticons are eliminated [13]. NLTK Python API was used in this task.

(3) **Location estimation:** This step is made of three phases: (1) Identifying all place names in textual descriptions, this is achieved using Google natural language API. When there is no location the text is discarded. (2) Disambiguation of the possible names for similar geographic locations, GeoNames Web Service is used to determine the granularity of the place. (3) A regional gazetteer identifies place names and synonyms within a particular area. (4) Finally, Google Maps Javascript API retrieves the coordinates of each location detected.

(4) **Traditional text classification:** Opinions, complaints, and impeachments are classified in one of the following classes: air, transport, traffic, health and water. The classes set is our social conceptualization of the air pollution phenomenon, we defined it with the advice of an air quality expert. The classification algorithm used is the Support Vector Machine (SVM). The SVM was selected for its geometric component [19, 20] that provides a tolerance band to indent entries during the process of categorizing opinions maintaining high performance in current benchmarks [21]. The dataset training has 1320 rows classified manually; texts were obtained from verified accounts, dataset allows the quality selection of social data. The performance in the testing phase was 87. SVM classifier uses a radial bases function kernel, parameters "c" (values from 1 to 10000) and "gamma" (values from 0.1 to 100), this process was implemented using Python GridSearch.

(5) **Text Clustering:** Each set of the classified publications is an input of the k-means clustering algorithm, this is a pipeline machine learning architecture (i.e., the output of the supervised method is the input of unsupervised method). The number of clusters generated lies from 5 to 10.

In Fig. 2 a summary of the detection of social geographic perception is shown. An extraction strategy is executed, for instance, a data set is collected where citizens perceive a relation of the pollution with high concentrations of traffic. Then the SVM detects several texts that belong to the class "Traffic"; they are the input of the K-means clustering [22]. Finally, the texts related to traffic are placed in 3 groups: "TrafficAlert", "Pyrotechnics" and "Protests". The first term is a hashtag used to alert traffic situations. The second term, after reviewing the grouped texts, we observed that citizens related pyrotechnics to the increase of air pollution. In the third term, social perception tends to relate traffic to the pollution problems. The question is "what is the spatio-temporal behavior of the social perception topics?".

3.2   Spatio-Temporal Mediation

We developed an air quality spatial data warehouse in Mexico City and part of State of Mexico (i.e., that contains data of over the last 30 years), other data sources like social protests and traffic are also included. We explore it through usual OLAP operators (e.g., Slice, RollUp, and DrillDown) obtaining detailed and aggregated outputs. Spatial cubes are described by numerical measures of air pollution at different spatio-temporal scales and granularities. For instance, the spatial hierarchy corresponds to municipalities and cities, while the time hierarchy to months, years and decades. Then, the perception dataset contains the original post, class, date, postal address detected (municipality, state, or place detected) and geographic coordinates.

The goal of the mediation is to detect spatial-temporal compatibility between social data perception and open data sources. The process consists of the following steps:

- Select spatial data cubes: According to the perception results of the spatial data mining question, we select or built the spatial cubes. For example, if traffic and social protest are trends according to the citizens, we choose cubes regarding of CO, NOX, PM10 y PM2.4.
- Define spatio-temporal scope: The records of each set of data are grouped per year or month, until reaching weeks only if there is a deep level of temporal granularity, then temporal hierarchies are created. Temporal coverage is estimated for each data set. Similarly, the data are grouped geographically, the available location columns are identified, rows are grouped according to each column, spatial hierarchies are created, these define the geographical coverage.
- Setting spatio-temporal values: The intersection between the temporal space coverage of each data set is performed, then matches are found. Therefore, we can define the values of the spatio-temporal equalizers that will help to adjust the data exploration in a unified way or for only one data source. The values for the temporal and spatial equalizer are summarized in Fig. 3.

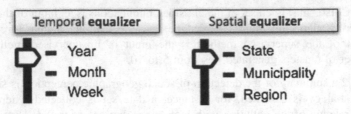

Fig. 3. Spatio-temporal equalizers

3.3 Multidimensional Analysis

This phase develops a cross-comparison of the data using open and social data cubes. Datasets have to match according to a spatio-temporal compatibility. In the opposite case it is not possible to analyze such data from a multidimensional approach, in fact this involves others open data related to the air pollution phenomenon (social protests, traffic, health, etc.). For instance: suppose that social perception indicates that social protests and air pollution are trends in the Gustavo Madero Region (GAM). When we explore a road incidents data cubes, a set of rows contains the location of "IPN avenue" at GAM (i.e., a municipality of Mexico City). However, in a traffic data cube, there are only rows located at GAM (i.e., it doesn't provide a specific location). How can we relate such data to different levels of geographic scales? How can we establish that GAM is a region within Mexico City? Using a spatial semantic hierarchy equivalence as illustrated in Fig. 4, the geographic scale can be managed using an adaptation of the algorithm introduced in [13] so-called OntoClassifier mediator (cf. Table 2). The semantic hierarchy is based on descriptions of places derived from a geographic domain (Mexico City and its municipalities).

Fig. 4. Spatial equivalence ontology

Table 2. OntoClassifier mediator algorithm

```
OntoClassifier Mediator Algorithm

1. Begin
2. Let q[i] name place
3. N = 0
4.  while n < i
5.   parsing and identification (q[i])
6.   node.start()
7.     while node != null
8.     j++, i++
9.     if similarity(concept_name)
          conVec[j]neighborhood_relations(node)
          node.next()
10.       geoVector[k] = geographic_semantic_search(conVec[j])
11.    geographic_thematic_search(conVec[j])
13.    j++, k++, n++
14. End
```

The OntoClassifier works as follows: the input is a place name extracted from the dataset. Therefore, it is correlated with the Ontology. If a match occurs, the context of the concept (i.e., all neighborhood concepts) is extracted, and it is stored into a vector (i.e., algorithm 1, lines 4 to 10). The vector is used to determine which domain and category the name place belongs; using the hyperonym relation of the ontology, we discover the domain and category. For example, according to the "IPN" place name "is an avenue", it belongs to "Gustavo A. Madero (GAM)", and "GAM" "is a place" of "Mexico City". Finally, the "social protest" and "road incidents" data cubes match in the GAM geographic region; then the OLAP/SQL statements are built to data exploration.

3.4 Data Visualization

Finally, we propose a user interface prototype to explore the unified data of the air pollution phenomena as revealed by our approach. Using the spatio-temporal controllers, one can discover hashtags, key terms, periods of time, places social and networks account names. This knowledge can be useful to identify new data sources. Figure 5 presents some user interface functions that provide a sort of observatory of human-geographic phenomena. The control functions applied to the different data sources group and show them whenever compatible in space and time.

Next, newspaper data show the social context at some given times when potentially exploring the data (in Fig. 5 the newspaper page shows a social protest on September 26, 2014). The user interface displays social perception and open data when combining maps and plots.

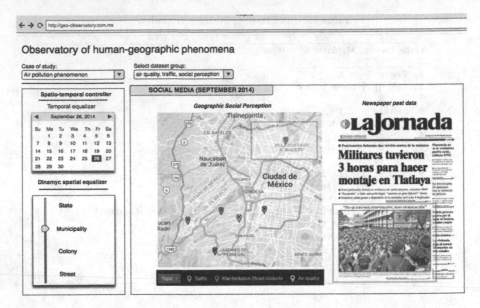

Fig. 5. UI prototype of human-geographic data visualization

4   Preliminary Experiments and Results

The experiments were executed in a semi-automatic way. We use the social perception dataset with 580,100 processed rows (Mexico City and State of Mexico, from 2014 to 2016), Air Quality open database that contains 33,120,000 rows (from 1986 to 2016), and road incidents (Mexico City, traffic, social protest, closed streets) with 20,000 rows (Mexico City, 2014).

According to the developed methodology, we explored several spatial big data questions provided by an air quality expert: What are the most important topics from society regarding air quality at Mexico City? Performing discovery of social perception, we detect more than 3000 posts classified in traffic. After a text clustering, the "pyrotechnics" term groups 449 texts, 1671 texts are grouped in "social protests" at Gustavo A. Madero (GAM) region. The perception analyzed was focused on September 2014. Those topics are trend at center and North of the city. From the social map illustrated in Fig. 6, protests and traffic appear as trend topics.

Last, an ultimate question was: Are there some historical relations between transport emissions and air quality? We built a spatial data cube of CO, NOX and PM elements related to transport emissions.

The size cube is 1,790,000 of rows, spatio-temporal time granularity: 20 years in Mexico City. Figure 7 shows that historically there is a bad quality at the North region of Mexico City on September 2014. Therefore GAM is one of the most polluted regions by car emissions.

Fig. 6. Geographical social perception (each mark groups more than 1000 posts)

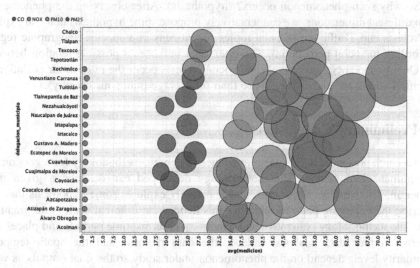

Fig. 7. CO, NOX and PM elements related to car emissions on September 2014 (Gustavo A. Madero (GAM) is one of the most polluted regions)

Finally, a final question was "Is there a temporal relation between road incidents and air quality?". We used a social protest cube applied to more than 8000 rows at Mexico City as illustrated in Fig. 8. We discovered that there are more than 200 social protests at GAM in 2014, as indicated by the hypothesis defined by social perception

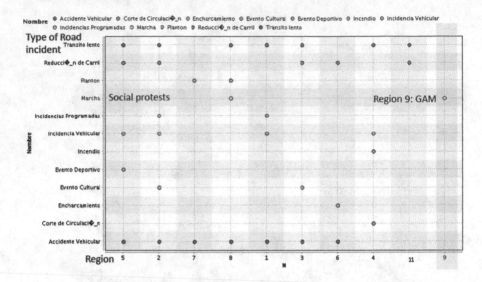

Fig. 8. Road incidents data visualization (social protests at GAM)

data. Also, there are 30 events on 2014 of closed streets due to sporting and cultural events that may affect traffic.

So why such phenomenon occurs? (hypothesis): After observing the phenomenon from different dimensions, we can tentatively propose some hypothesis that can explain it: Protests and Traffic are relevant topics for citizens in a specific geographic region (according to social perception). The open data indicates a possible relation between Air Quality and social protests. When social protests occur the rates of air pollution at specific regions are higher (i.e., more than 60 average points historically).

5 Preliminary Conclusions

The research presented in this paper introduces a generic methodology for a geographic and human phenomenon. The methodology developed can describe a given phenomenon, discover spatial patterns, and tentatively explores some hypothesis that can describe the emerging patterns and why this happens at different scales and granularities. The methodology is iterative and can discover terms, time periods and places that improve further data analysis. Indeed, the data richness and the spatio-temporal granularity levels depend on the phenomenon under study so the final outputs as well.

According to the preliminary experiments the geographical social perception complements the understanding of the phenomena under study. Therefore, social media posts about general air pollution topics establish a general level of geographic granularity, opposite to complaints (i.e., regarding a specific topic) that establishes specific locations. Future work will consider additional data analysis and experiments, deep learning methods on additional open data sources and integration of other spatio-temporal approaches, as well as developing advanced data visualization interfaces. In particular, we plan to include a dataset of air quality derived from satellite images, this will contribute to increase the variety of information, and would improve the veracity or discard of the conclusions obtained from the social perception. This analysis would imply a challenge in the speed of the flow of data, because satellite data is generated periodically, increasing the information sampling to be analyzed in real time.

Acknowledgment. The authors of this paper thank God, CONACYT project number 1051, the Laboratorio de Cómputo Móvil-UPIITA, COFAA-IPN, SIP-IPN project 20171086, and Instituto Politécnico Nacional (IPN) for their support.

References

1. Hardy, K., Maurushat, A.: Opening up government data for Big Data analysis and public benefit. Comput. Law Secur. Rev. **33**(1), 30–37 (2017). https://doi.org/10.1016/j.clsr.2016.11.003. ISSN 0267-3649
2. Chen, X., Shao, S., Tian, Z., Xie, Z., Yin, P.: Impacts of air pollution and its spatial spillover effect on public health based on China's big data sample. J. Clean. Prod. **142**, 915–925 (2017). https://doi.org/10.1016/j.jclepro.2016.02.119. ISSN 0959-6526
3. Ang, L., Phooi, K.: Big sensor data applications in urban environments. Big Data Res. **4**, 1–12 (2016). https://doi.org/10.1016/j.bdr.2015.12.003. ISSN 2214-5796
4. Mayer-Schonberger, V., Cukier, K.: Big Data: A Revolution That Will Transform How We Live, Work, and Think. Eamon Dolan/Houghton, Mifflin Harcourt (2014)
5. Marien M. Global challenges for humanity (2014). http://www.millennium-project.org/millennium/challenges.html
6. Miller, H., Han, J.: Geographic Data Mining and Knowledge Discovery. Chapman & Hall/CRC Data Mining and Knowledge Discovery Series, 2nd edn. CRC Press, Boca Raton (2007)
7. Chakrabarti, A.: Cross-correlation patterns in social opinion formation with sequential data. Phys. A Stat. Mech. Its Appl. **462**, 442–454 (2016). ISSN 0378-4371
8. Bakshy, E., Messing, S., Adamic, L.: Exposure to ideologically diverse news and opinion on Facebook. Science **348**, 1130–1132 (2015). Sciencemag.org
9. Lee, J., Kang, M.: Geospatial big data: challenges and opportunities. Big Data Res. **2**(2), 74–81 (2015). https://doi.org/10.1016/j.bdr.2015.01.003. ISSN 2214-5796
10. Blazquez, D., Domenech, J.: Big Data sources and methods for social and economic analyses. Technol. Forecast. Soc. Change (2017). https://doi.org/10.1016/j.techfore.2017.07.027. ISSN 0040-1625
11. Zagal-Flores, R., Mata, M., Claramunt, C.: Geographical knowledge discovery applied to the social perception of pollution in the City of Mexico. In: 9th ACM SIGSPATIAL International Workshop on Location-Based Social Networks (2016). https://doi.org/10.1145/3021304.3021307
12. Mata, F., et al.: A mobile information system based on crowd-sensed and official crime data for finding safe routes: a case study of Mexico City. Mob. Inf. Syst. 11 p. (2016). https://doi.org/10.1155/2016/806. Article ID 8068209
13. Zagal-Flores, R., Mata-Rivera, F., Claramunt, C.: Discovering geographical patterns of crime localization in Mexico City. In: WEB 2017: The Fifth International Conference on Building and Exploring Web Based Environments (2017). ISBN 978-1-61208-557-9

14. Mata, F., Torres-Ruiz, M., Zagal, R.: A cross-domain framework for designing healthcare mobile applications mining social networks to generate recommendations of training and nutrition planning. Telematics Inform. (2017). https://doi.org/10.1016/j.tele.2017.04.005

15. Yuan, M.: Use of knowledge acquisition to build wildfire representation in geographic information systems. Int. J. Geogr. Inf. Syst. **11**, 723–745 (1997)

16. Di Martino, S., Bimonte, S., Bertolotto, M., Ferrucci, F., Leano, V.: Spatial online analytical processing of geographic data through the Google earth interface. In: Murgante, B., Borruso, G., Lapucci, A. (eds.) Geocomputation, Sustainability and Environmental Planning. Studies in Computational Intelligence, vol. 348. Springer, Heidelberg (2011). https://doi.org/10.1007/978-3-642-19733-8_10

17. Mahboubi, H., et al.: Semi-automatic design of spatial data cubes from simulation model results. Int. J. Data Warehous. Min. **9**(1), 70–95 (2013). Academic OneFile. http://link.galegroup.com/apps/doc/A340297894/AONE?u=pu&sid=AONE&xid=01ca4c69

18. Wakamiya, S., Belouaer, L., Brosset, D., Kawai, Y., Claramunt, C., Sumiya, K.: Exploring geographical crowd's emotions with Twitter. Inf. Media Technol. (2015). https://doi.org/10.11185/imt.10.35. Online ISSN 1881-0896

19. Bishop, C.: Pattern Recognition and Machine Learning. Springer, New York (2006). ISBN 978-0-387-31073-2

20. Abu-Mostafa, Y., Magdon-Ismail, M., Lin, H.: Learning from Data: A Short Course (2012). AMLBOOK.com

21. Zhang, C., Liu, C., Zhang, X., Almpanidis, G.: An up-to-date comparison of state-of-the-art classification algorithms. Expert Syst. Appl. **82**, 129–150 (2017). ISSN 0957-4174

22. Srivastava, A., Text, S.M.: Mining: Classification, Clustering, and Applications. CRC Press, Boca Raton (2009). 328 pages

23. Zheng, Y., Chen, X., Jin, Q., Chen, Y., Qu, X., Liu, X., Chang, E., Ma, W., Rui, Y., Sun, W.: A cloud-based knowledge discovery system for monitoring fine-grained air quality. MSR-TR-2014-40. Microsoft Research Asia (2014)

24. Zheng, Y., Yi, X., Li, M., Li, R., Shan, Z., Chang, E., Li, T.: Forecasting fine-grained air quality based on Big Data. In: Proceedings of the 21th ACM SIGKDD International Conference on Knowledge Discovery and Data Mining (KDD 2015), pp. 2267–2276. ACM, New York (2015). https://doi.org/10.1145/2783258.2788573

25. Chen, X., Shao, S., Tian, Z., Xie, Z., Peng, Y.: Impacts of air pollution and its spatial spillover effect on public health based on China's big data sample. J. Clean. Prod. **142**(Part 2), 915–925 (2017). ISSN 0959-6526

A Web of Data Platform for Mineral Intelligence Capacity Analysis (MICA)

Danielle Ziébelin[1](✉), Philippe Genoud[1], Marie-Jeanne Natete[1],
Daniel Cassard[2], and François Tertre[2]

[1] Université Grenoble-Alpes CNRS LIG, IMAG building - 700 avenue Centrale
Domaine Universitaire, 38401 Grenoble, St Martin d'Hères, France
{danielle.ziebelin,philippe.genoud,
marie-jeanne.natete}@imag.fr
[2] BRGM, 3, avenue Claude Guillemin, BP 36009
45060 Orléans Cedex 2, France
{d.cassard,F.Tertre}@brgm.fr
http://www.liglab.fr/fr/presentation/equipes/steamer
http://www.brgm.fr

Abstract. The H2020 Mineral Intelligence Capacity Analysis (MICA) project is
a collaboration between multiple geological surveys (BRGM, BGS, GEUS,
GeoZS) in Europe; the European Raw Materials Intelligence Capacity Platform
(EU-RMICP) enables the user to find information about mineral raw material in
Europe. This paper discusses a part of EU-RMICP platform based on the models
and the Web of data architecture. This part of the platform focusing on mineral
resources, methods and information collected in different European data-bases, to
leverage semantic technologies and to manage and link geoscience information
and resources. Experts deal with many representations by taking into account the
environmental, technical, political and social dimensions, metadata, heterogeneous data sources and tools. Our solution for semantic interoperability between
different resources, is based on semantic annotation by adding knowledge to
resources with semantic tags. The semantics attached to resources is defined by
ontology and the exploitation is made with an integrated model, which maps the
correspondence between the ontology and the resources. The ontology is based
on an ontology of the domain of mineral resources coupled with relative commodities, time and space. The study focuses on the topic of semantic modeling,
exploratory data, ranking the queries results and the query of linked data are
introduced through Euro-Lex datasets. The architecture is based on a RDF triple
store storing the ontology, the methods and documentation, the scenarios and
metadata. The triple store is connected with eight existing European Data bases,
and with the inference engine to search, select, infer and rank the results.

Keywords: Web of data · Semantic geospatial web and interoperability
Geospatial databases and spatio-temporal data management

1 Introduction

Researchers in various disciplines of geoscience have developed explicit knowledge
from their observation, datasets, the design of methods and procedures, data structures,
data documentation. These identified concepts are articulated and defined in a

© Springer International Publishing AG, part of Springer Nature 2018
M. R. Luaces and F. Karimipour (Eds.): W2GIS 2018, LNCS 10819, pp. 155–171, 2018.
https://doi.org/10.1007/978-3-319-90053-7_15

knowledge base, and specify the domain and subjects of discipline and are used to develop models. This knowledge is also, the start point for data reuse and data inter-operability especially when datasets are collected from different sources. The Semantic Web technologies record this knowledge and build knowledge bases to underpin data resources and make it accessible and readable for both humans and machines. Building an ontology is one possibility to access this knowledge. Each ontology is the formal specification of a shared conceptualization of a domain [1]. W3C develops many standards to guide and formalize the modeling and encoding of ontologies, as well as Linked Open Data. The data structure in the Semantic Web is the Resource Description Framework (RDF, https://www.w3.org/TR/rdf11-primer/) which has a triple form "Subject, Predicate, and Object".

In the field of geoscience, ontologies and knowledge bases are multiple: upper ontology domains like GEON [2] or SWEET [3], domain ontologies in tectonic, rock classification, hydrology, minerals, ... or application ontologies based on data format and metadata description. These formalizations increase the potential indexation, interoperability and publication of collected datasets. The Linked Open Data approach in geosciences is based on OGC and W3C standards building discipline data models, ontologies and vocabularies, the challenge is to coordinate the data standards from different disciplines and develop efficient methods to implement them such as GeoS-ciML (http://www.geosciml.org) and OGC standard-based data services WMS. The web of data and the Linked Open Data (LOD) adds annotations and identifications of resources on the Web and facilitate linkages between resources. The Web of Data gives opportunities for exploring more information about geosciences and crowd-sourced datasets.

This MICA project shows experimental results with several European data sources; for domain specific ontology and vocabulary a specific knowledge-base is made based on mineral ontology. The linked data is made by queries on internal and external resources with SPARQL language (https://www.w3.org/TR/sparql11-overview).

2   Project Presentation

The MICA platform was developed in the frame of the H2020 MICA project; the description of the project can be found on the project website (www.mica-project.eu/) [4]. To briefly summarize: MICA (Mineral Intelligence Capacity Analysis 2015-2018) objectives are to develop a platform of knowledge, the EU-Raw Materials Intelligence Capacity Platform (or EU-RMICP), integrating metadata on data sources related to primary and secondary mineral resources and bringing the end-users expertise on the methods and tools used in mineral intelligence. The EU-RMICP is based on semantic web technology with an ontology of the domain of mineral resources coupled with cross-functional ontologies, relative to commodities, time and space. The system is coupled with an 'RDF Triple Store' which indexes resources: sheets (i.e., specific formatted forms) related to methods and documentation, metadata and existing data bases: EURare www.eurare.eu/, Minerals4EU www.minerals4eu.eu/, ProSUM www.prosumproject.eu/ [5], SCRREEN http://scrreen.eu/, EGDI http://www.europe-geology.eu/ and the RMIS 2.0 http://rmis.jrc.ec.europa.eu/ (Fig. 1).

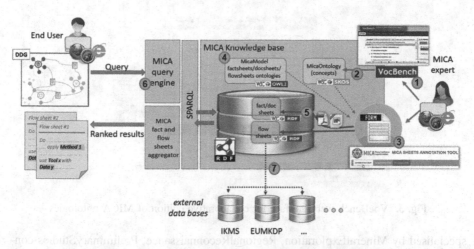

Fig. 1. General architecture of the MICA Materials Intelligence Capacity Platform. (1) Voc-Bench Ontology Editor to build the MICA Multidimensional Ontology, (2) to create SKOS Multidimensional Ontology, (3) MICA sheets Annotation tool, (4) Model of Sheets in OWL, (5) RDF graph, (6) MICA Query Engine, (7) to annotate external databases and resources.

3   The MICA Main Ontology and the Associated Knowledge

The MICA Ontology covers 7 thematic domains: 'Primary' and 'Secondary Mineral Resources', 'Industrial Processing and Transformation', 'Raw Materials economics' (including CRMs), 'Raw materials Policy & Legal Framework', 'Sustainability of Raw Materials' and 'International Reporting'. To define the MICA ontology we use SKOS vocabulary [https://www.w3.org/2004/02/skos/]. Each concept is described with a SKOS predefined property. Figure 2 shows the main classes and properties defined in the MICA Multidimensional Ontology for representing Main Ontology: D1 Primary resources and D6 Sustainability of raw materials. Each concept is described with a SKOS predefined property (skos:prefLabel, skos:definition, etc.), skos:broader properties connecting concepts and sub-concepts. etc.). skos:broader properties connect concepts and sub-concepts (e.g., in Fig. 2, D1 PrimaryResources has a sub-concept which is

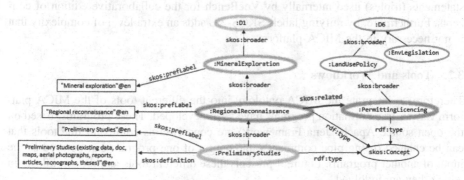

Fig. 2. SKOS ontology implementation.

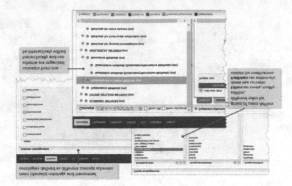

Fig. 3. VocBench web-interface for collaborative edition of MICA ontologies

specialised by MineralExploration, RegionalReconnaissance, PreliminaryStudies concepts). The skos:related property is used to link two concepts that are not hierarchically related (e.g., RegionalReconnaissance with PermittingLicensing).

3.1 Ontology Management

To facilitate the construction of the MICA Ontology (main and transversal ontologies), VocBench editor [5] has been chosen (Fig. 3). VocBench is a Web-based, multilingual, collaborative ontology editor developed by the ART Research Group at the University of Rome Tor Vergata. VocBench manages OWL ontologies, SKOS-XL thesauri and generic RDF datasets. VocBench is used by the different partners to collaboratively edit the MICA Ontology. Various accounts have been created for the different partners with different roles: editors, who can create/update/delete concepts; publishers, who can additionally approve or reject the modifications proposed by the editors and readers, just for consultation.

The MICA Ontology created with VocBench uses the SKOS-XL www.w3.org/2008/05/skos-xl, vocabulary to represent MICA concepts and the relationships between them. It can be exported as an RDF file in various RDF serialization formats (RDF/XML, Turtle, NTriples). The exported ontology is not directly usable by the various tools developed in the MICA platform. In particular, it contains extra RDF statements (triples) used internally by VocBench for the collaborative edition of concepts. Furthermore, by reifying labels SKOS-XL adds an extra level of complexity that is not necessary in the MICA platform.

3.2 Tools and Workflows

Therefore, to integrate the MICA ontology into the different tools of the MICA platform various transformation programs have been developed. These programs (based on the open source Apache Jena Framework) are conceived as command line tools that can be chained (using pipe commands, the output of one program can be used as the input of another program). Figure 4 presents these different tools and the workflows in which they are involved.

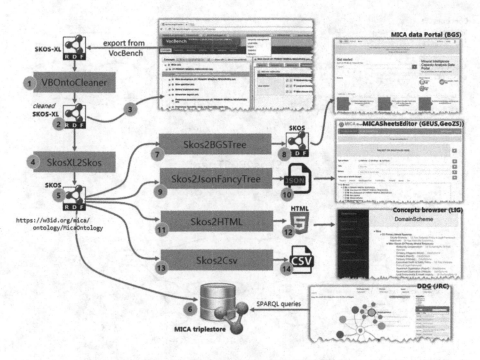

Fig. 4. Management tools for the MICA ontology

The VBOntoCleaner – 1- program takes as input the SKOS-XL file generated from an export form of VocBench. It removes all extra statements internally used by Voc-Bench and definitively deletes the concepts whose suppression has been approved by MICA expert publishers. It produces a new SKOS-XL file -2- that can be re-imported into VocBench -3- as a new MICA ontology version for a future collaborative edition or used as input of the chain of treatments that need to be performed in order for it to be used by the MICA platform tools. The SkosXL2Skos -4- program transforms the SKOS-XL file into a regular SKOS file -5- (it transforms reified labels of SKOS-XL into simple string literals). This SKOS file corresponds to the MICA Ontology published with URI https://w3id.org/mica/ontology/MicaOntology. This Ontology is the one used by the MICA platform to annotate the MICA resources in the MICA database -6- and by the web interface to retrieve these resources. In the MICA SKOS Ontology, the concepts are organized using skos:ConceptsSchemes. Each concept scheme regroups the concepts either of the main ontology or of one of the transversal ontologies (data, methods, tem-poral, spatial …). In each concept scheme, concepts are organized in hierarchies, with top-level concepts directly attached to the scheme. This structure is not practical for the GeoNetwork catalog for MICA data sources that needs a "pure" tree hierarchy of concepts. The Skos2BGSTree program -7- transforms the MICA Ontology SKOS file into such a tree of concepts (Fig. 5). It adds two "artificial" levels of concepts: a top-level concept and sub-concepts that correspond to the concepts schemes. The output is a RDF/XML SKOS file -8- that can be used by the GeoNetwork catalog.

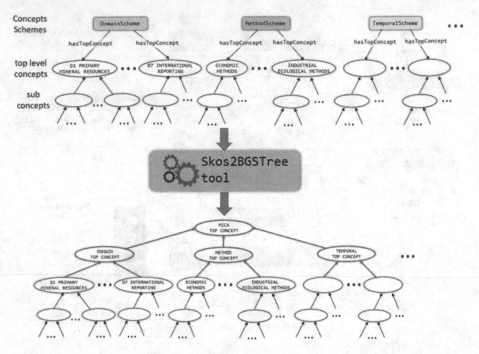

Fig. 5. Transformation of MICA main (Domain) ontology and transversal ontologies structured with concepts schemes into a "pure" tree of concepts.

In the same way, the Skos2JsonFancyTree program -9- transforms the MICA Ontology SKOS file into a JSON file -10- that can be directly used by the FancyTree JavaScript component that displays the concept hierarchy in the MICASheetEditor. VocBench is well suited for editing the MICA Ontology, but it is less convenient for browsing all the trees of concepts and, for a given concept, to visualize immediately the related concepts. In order to facilitate this navigation for MICA experts elaborating the ontologies, two programs have been developed:

- Skos2HTML -11- that takes the MICA Ontology SKOS/RDF file as input and produces an HTML page -12-, see Fig. 6 (http://lig-coin.imag.fr/mica/concepts.html).
- Skos2CSV -13- that takes the MICA Ontology SKOS/RDF file as input and produces a CSV file -14-.

4   MICA Data Model

The MICA Model ontology defines a data model for the MICA project. This data model is implemented as an OWL ontology that defines the vocabulary used for describing the MICA resources stored in the database (triple store) and annotated with the concepts defined by the MICA Ontology. The main classes of this model are described below. MICAResource: A MICA Resource is any resource defined in MICA database (triple store) which is annotated with MICA concepts (defined in the MICA Ontology). Resources can be sheets, questions, link to external resources etc. A resource can be linked to other resources using the relatedTo property. MICAQuestion: A MICA Question is any sentence which asks about the MICA project and that can be an-swered by some MICA Knowledge Elements. MICAKnowledgeElement: A MICA Knowledge Element is a piece of knowledge concerning Mineral Intelligence that has been identified by MICA experts and has been annotated with MICA concepts defined in the MICA Ontology. The expert has also to give it a title and a small summary.

MICASheet: A Sheet is an elementary piece of knowledge (MICAKnowl-edgeElement). This knowledge is accessible through a URI (public URI). This URI can identify either a resource already available on the Web (for example legislation from EUR-Lex) or a new resource that MICA experts have produced. The public URI allows to access a representation of the resource. MICAContentType: defines the various content types a sheet can have (Methods&Tools, Documentation …). LinkedSheet: for the knowledge comes from a resource external to the MICA platform and is accessible on the Web (through its publicURI). FlowSheet: A flowSheet is an ordered list of MICA Knowledge Elements typically, a flowSheet can be seen as a "recipe" for answering a MICA Question. A full description of this ontology (classes and proper-ties) is available online and can be accessed through the ontology URI: https://w3id.org/mica/ontology/MicaModel (Fig. 7).

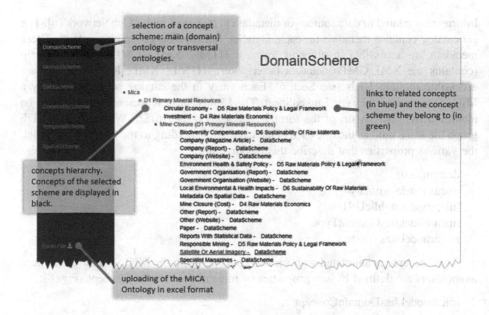

Fig. 6. HTML page for rapid navigation between related concepts.

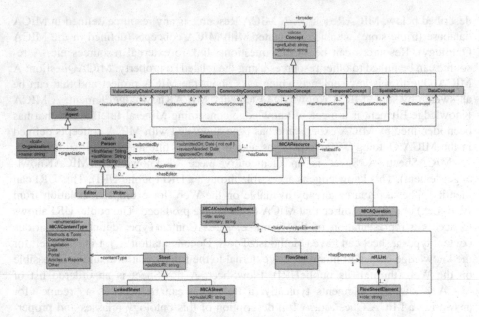

Fig. 7. UML class diagram view of the classes and properties defined in this data model.

5   Relationships with GeoNetwork

Information related to data sources or metadata is managed under GeoNetwork [6]. The geocatalog contains metadata for each: DataSet, Series, Non geographic dataset. These metadata are accessible in different formats: HTML, XML (GMD), RDF, pdf, zip (contains the XML:GMD metadata description). HTML, XML, RDF metadata are accessible through URIs (see Sect. 6). Each entry in the catalog is represented by a MICAResource of the type LinkedSheet. The LinkedSheet representing a data will be identified by a MICA uri of the form `micaresource:uuid`. We use in the RDF graph the same uuid as the one used in the catalog. According to the MICA data model, the various properties that describe this resource are:

> dcterms:title
> micamodel:summary
> micamodel:publicURI
> micamodel:hasContentType
> micamodel:hasWriter

This resource is annotated with various concepts from the MICA ontologies. These annotations are defined by sub-properties of micamodel:hasMicaConcept (Fig. 8):

> micamodel:hasDomainConcept
> micamodel:hasCommodityConcept
> micamodel:hasDataConcept
> micamodel:hasTemporalConcept
> micamodel:hasSpatialConcept

micamodel:hasMethodConcept
micamodel:hasSupplyChainConcept

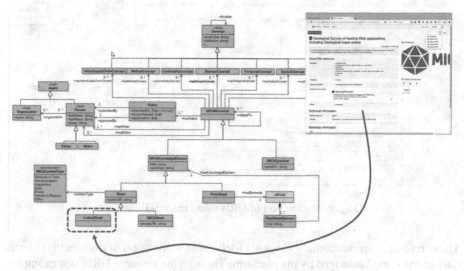

Fig. 8. Link between geonetwork and Mica Data Model.

To define a mapping between metadata and those properties, and then to define the operations to create/update/delete this information inside the MICA triple store the system provide algorithms of mapping.

- For example the GeoNetwork uses other descriptive keywords coming from others ontologies (for example GEMET). To generate the annotations in the MICA triple store it is necessary to use an algorithm which parses the concept scheme information present in the GMD/XML file using a mapping between MICA concepts URIs and the concept scheme present in the GMD/XML file.
- Insertion of triples into the MICA triple store is performed through SPARQL1.1 update queries. The program CreateLinkedSheet realizes such a query with a Java program using Jena API.
- If the resource is already present in the triple store, it may be necessary to update the database by adding new triples, replacing existing ones or removing some existing triples, by removing all the existing triples involving the resource as subject, and inserting all the triples that have been extracted for the GMD/XML file as if the resource was a new one (same as case 1).

6   MICA URIs

6.1   Persistent URIs

According to the Web of Data principles, all resources in the MICA platform are identified using HTTP URIs (Uniform Resource Identifiers) as persistent URIs [7].

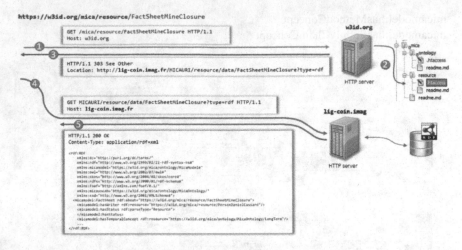

Fig. 9. Redirection of URIs using the w3id service.

These resources are vocabulary elements (defined in ontologies) or documents (Sheets, data sources, etc.) managed by the platform. Through the resource URIs (for example a MICASheet) it is possible to access various representations: a PDF document, a docx document, and an RDF representation, etc. For this reason it is important to choose URIs for MICA resources in a domain whose HTTP server can be configured to either serve the vocabulary elements directly, or host a resolver or redirection service. We chose to use https://w3id.org, to define a MICA domain, and to obtain persistent URIs. Proceeding this way, all the URIs concerning MICA resources start with the same prefix: https://w3id.org/mica/. This w3id.org service is run by the W3C Permanent Identifier Community Group, and provides a secure, permanent URL re-direction service for Web applications. This service uses Apache rewriting rules expressed in the access files (Fig. 9).

6.2 MICA Namespaces

As we saw above, HTTP URIs (Uniform Resource Identifiers) are used to uniquely identify the entities used in the MICA project, and we use persistent URIs based on the w3id.org service. So all MICA URIs are defined in namespaces starting with the same string: https://w3id.org/mica/. Three namespaces are defined:

- https://w3id.org/mica/ontology/MicaModel#: the prefix used for all the classes and properties described in the OWL ontology defining the data model for the resources managed by the MICA platform.
- https://w3id.org/mica/ontology/MicaOntology/: the prefix used for all the SKOS concepts for the MICA domain and transversal ontologies.
- https://w3id.org/mica/resource/: the prefix for all the resources managed by the MICA platform and for which an RDF representation conforming to the MICA data model is stored in the MICA Triple Store.

For the MICA Model ontology the local name part of the URIs is the class or property name, for example:

- https://w3id.org/mica/ontology/MicaModel#Sheet for the Sheet class,
- https://w3id.org/mica/ontology/MicaModel#hasDomainConcept for the hasDomain Concept property. For concepts defined in the MICA Ontology, URIs are of the form https://w3id.org/mica/ontology/MicaOntology/xxxxx where xxxxx is a random UUID. For example, https://w3id.org/mica/ontology/MicaOntology/fe14c 339024c48a893fb7ec2a322071c is the URI of the Carbon Footprint concept. The same rule applies for resources created by the MICASheetEditor and stored in the MICA triple store: the local name part of the URI is a UUID. An example of such a URI: https://w3id.org/mica/resource/71b5acd42e5447d491460e9dd6dbd30

6.3   Dereferencing MICA URIs

The Linked Data Third principle states that "When someone looks up a URI, provide useful information, using the standards", in other words, URIs should be dereferenceable, meaning that HTTP clients can look up the URI using the HTTP protocol and retrieve a description of the resource that is identified by the URI. Given a resource URI, any client should be able to retrieve a representation of the resource in a form that meets his/her needs, such as HTML or PDF for humans and RDF for machines.

The HTTP mechanism called content negotiation addresses this issue (Fig. 10). The basic idea of content negotiation is that HTTP clients send HTTP headers with each request to indicate what kind of documents they prefer. Servers can inspect these headers and select an appropriate response. For example, if the headers indicate that the client prefers HTML, then the server will respond by sending an HTML document. If the client prefers RDF, then the server will send the client an RDF document (which may be in various serialization formats).

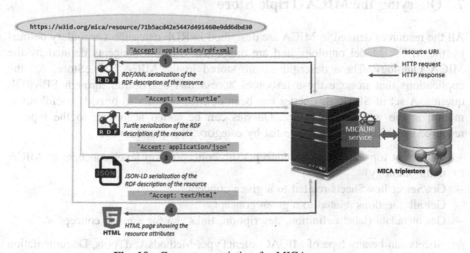

Fig. 10. Content negotiation for MICA resources

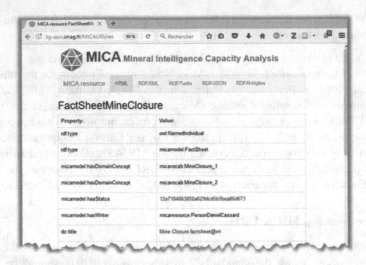

Fig. 11. HTML page corresponding to the MICASheet resource about Mine closure.

We adopted this approach for all MICA URI's. For example, given a MICA resource URI identifying a resource stored in the MICA triple store, it is possible to obtain its RDF description in a given RDF serialization format (RDF/XML, Turtle or JSON-LD) or as a HTML page (see Fig. 11). The representation format is defined in the Accept: header of the HTTP request used to dereference the resource URI. When PDF files for MICA sheets are stored in the MICA database, the MICA-URI service implements this content negotiation and allows direct retrieval of these documents using the resource URI (for example for visualization in the web user interface).

7   Querying the MICA Triple Store

All the resources defined in MICA are described in RDF using the vocabulary defined by the MICA Model ontology and are annotated with the concepts defined by the MICA Ontology. These descriptions are stored in the MICA TripleStore. All the applications that manage these resources access this description through SPARQL queries. A set of SPARQL queries has been defined in order to provide useful information in the web user interface. Queries can be sorted according to the type of resource to which they relate. We list by category, the queries made:

– Get/Select top level domain concepts/sub-concepts (top level concepts in MICA main ontology)
– Get/Select flowSheets related to a given concept
– Get all questions related to a given concept
– Get metadata (label definition, description, links …) for a given concept

Sheets can be any type of MICAContentType, MethodsAndTools, Documentation and Legislation …: Get/Select all sheets of a given type of content; Get metadata (label definition, description, links ...) for a given sheet; Get statistics by sheetAnnotating MICA resources

The MICASheetEditor annotates resources (MICASheets, linkedSheets, flow-Sheets) using MICA concepts and stores these annotations in the MICA Triple Store. To create these annotations, the MICASheetEditor uses SPARQL 1.1 Update queries that permit CRUD (Create, Read, Update and Delete) operations on the graph data base. Queries can be sorted according to the type of resource to which they relate:

- MICASheets Queries concern MICASheets
- linkedSheets Queries concern linkedSheets
- MICAQuestions Queries concern MICAQuestions
- flowSheets Queries concern flowSheets
- RelatedResources Queries concern relations between MICAResources (skos:re-latedTo relationship).

The SPARQL queries performed on the whole graph, these interactions result which is composed of RDF representations of sheets, and MICA and Mica Model ontologies (extended by the inference mechanism of RDFs and OWL DL).

For example, *retrieve all Sheets about "Mining Wastes" with Sheets about related methods* corresponds to following SPARQL query:

```
SELECT DISTINCT ?fsd ?d ?fsm ?meth
    WHERE {   {   ?fsd model:hasDomainConcept ?d.
        FILTER(        ?d        =        micavocab:        Mining_Wastes        ||
            EXISTS {?d skos:broaderTransitive micavocab:Mining_Wastes}) }
    OPTIONAL {    {  ?m skos:inScheme micavocab:MethodsScheme;
            skos:related ?d.
        {   ?fsm model:hasMethodConcept ?m.
            BIND (?m as ?meth)        }
        UNION {
            ?m1 skos:broaderTransitive ?m.
            ?fsm model:hasMethodConcept ?m1.
            BIND (?m1 as ?meth)        }    }
    UNION {
        ?d1 skos:broaderTransitive ?d.
        ?m  skos:inScheme micavocab:MethodsScheme;
            skos:related ?d1.
        {   ?fsm model:hasMethodConcept ?m.
            BIND (?m as ?meth)        }
        UNION {
            ?m1 skos:broaderTransitive ?m.
            ?fsm model:hasMethodConcept ?m1.
            BIND (?m1 as ?meth)        }    }  }
    } ORDER by ?fsd ?fsm
```

The results of this query are shown on Fig. 12.

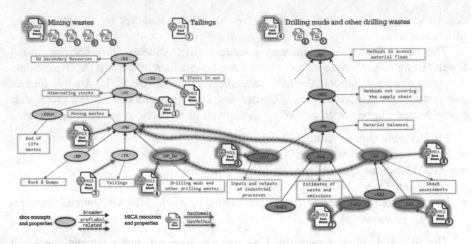

Fig. 12. Using the concepts hierarchy to retrieve sheets and relevant methods.

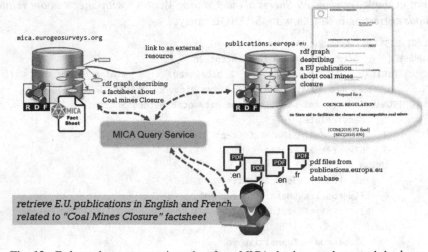

Fig. 13. Federated query to retrieve data form MICA database and external databases

Queries to external data sources: It is possible, to connect the RDF representation of a Sheet to external resources also described with RDF. This was experimented by connecting some factsheets to the E.U. law and publications database, and then performing some SPARQL federated queries (Fig. 13).

7.1   The Ranking of Results by Relevance

In order to make the system more powerful and more convenient for the end-user, the results are presented by pertinence or relevance. This will allow the MICA resources presented to the end-user to be ranked. These different resources (MICASheets for documentation, methods and tools, articles and reports, flowSheets for complex scenarios, linkedSheets for 'external high-quality' resources, etc.) are attached to one or several concepts from the MICA Ontology. They can also be linked together: for example, a MICASheet can be linked, to one or several sheets that detail some aspects (i.e., some piece of EU legislation, some types of data, etc.).

To illustrate this, let us consider the following example. Suppose we have in the MICA Triple Store five sheets annotated with D1 PRIMARY MINERAL RESOURCES concepts as shown in Fig. 11. We also have:

- R1 with concept C1_1, Resource assessment (sub-concept of C1 Mineral Exploration)
- R2 with concept C1_1_1, Approximate resource calculation (sub-concept of C1_1)
- R3 with concept C1_1_3, Geological Interpretation (sub-concept of C1_1)
- R4 with concepts C1_1_2, Drilling Assessment, and C1_1_3, Geological Interpretation, (both sub-concepts of C1_1)
- R5 with concept C1_2, Subsurface Exploration, (sub-concept of C1).

Ranking Task: the results must be ordered according to their relevance. The relevance is based on a combination of the semantic distance between concepts annotating a resource and the concepts used to express the search and the number of concepts annotating the resource. The semantic distance between two hierarchically related concepts is the number of skos:broader edges that separate the concepts. For example the semantic distance between concepts $C\_1\_1\_2\_1$ (Percussion Drilling Assessment) and C_1_1 (Resource Assessment) is 3 as shown in Fig. 14. To rank the resources, priority is given to the ones which are annotated with concepts that are closer to the concepts used to define the search. Thus, the ranking algorithm for a single search (search based on only one concept) consists, for each retrieved resource, of calculating:

(1) minDist, the minimum distance between concepts annotating the resource and the search concept. Formally, for a resource annotated with concepts c_i, $i \in [1..n]$, minDist = min(semanticDistance(c_i,c) where c is the searched concept.

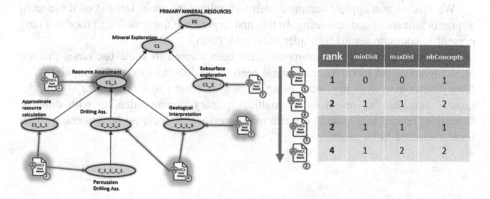

rank	minDist	maxDist	nbConcepts
1	0	0	1
2	1	1	2
2	1	1	1
4	1	2	2

Fig. 14. Ranking of resources for search based on C1_1 (Resource Assessment) concept.

(2) maxDist, the maximum distance between concepts annotating the resource and the search concept. Formally, for a resource annotated with concepts c_i, $i \in [1..n]$, maxDist = max(semanticDistance(c_i,c) where c is the searched concept.

(3) nbConcepts the number of concepts that annotate the resource.

The first ranking criteria is minDist. Resources are ordered with an ascending minDist value (resource with the smallest minDist is the first and so on …). In the case of equality, the second criteria is based on maxDist in an ascending order (resource with the small-er maxDist will be the first and so on …). Finally, in the case of equality on minDist and maxDist criteria, the nbConcepts criteria in a descending order is used (resource with the biggest nbConcepts criteria is the first and so on …).

8   Conclusion

After 26 months of the project, there has been some progress in the modelization and interoperability of different sources and data-bases. The conceptualization of the mineral intelligence has been significantly improved with full support for geological survey consensus concerning the concepts and the relations expressed in the ontology.

The performance of a semantic web framework has been improved for semantic interoperability and linked data in mineral intelligence.

The platform we have developed supports the interoperability of multiple hetero-geneous resources which are harvested automatically. The RDF graph has been created and used with several datasets. Existing standards such as INSPIRE and NUTS (Nomenclature of Territorial Units for Statistics) and time representation ISO 8601 have been used.

All of these components, validated in a proof of concept, will be refined and mature in future project. This work has opened various perspectives concerning what remains to be done in order to develop a complete framework for heterogeneous integration. The annotation processes needs to be refined with an identification and a characteri-zation of semantic annotation according to the types of resources (unstructured, semi-structured, structured) and to provide rules and recommendation in an annotation tool.

We studied and applied semantic web technology the main interest of these tech-niques is inferences and reasoning. In this first experience the reasoning process is only present to compute results to a query and rank them.

The ontology and inference model have been chosen to some use cases that we intend to solve; an intensive tests phase need to be done to extend this capability. In theory describing data resources using ontologies means that ontologies could ade-quately annotate this resources; in multidisciplinary domains dealing with different ontologies related to a specific domain may contribute to increase heterogeneity. That issue also need to be evaluate.

References

1. Gruber, T.: Toward principles for the design of ontologies used for knowledge sharing. Int. J. Hum. Comput. Stud. **43**, 907–928 (1992)
2. Lin, K., Lundascher, B.: GEON: ontology-enabled map integration. In: 24th Annual ESRI International User Conference (2004)
3. DiGiuseppe, N., Pouchard, L.C., Noy, N.F.: SWEET ontology coverage for earth system sciences. Earth Sci. Inf. **7**(4), 249–264 (2014). https://doi.org/10.1007/s12145-013-0143-1
4. Machacek, E., Falck, W.E., Delfini, C., Erdmann, L., Petavratzi, E., Van der Voet, E., Cassard, D.: Clearing the sky from the clouds - the Mineral Intelligence Capacity Analysis (MICA) project. Eur. Geolo. J. **44**, 48–53 (2017)
5. Huisman, J., Habib, H., Guzman Brechu, M., Downes, S., Herreras, L., Løvik, A.N., Wäger, P., Cassard, D., Tertre, F., Mählitz, P., Rotter, S., Chancerel, P., Ljunggren Söderman, M.: ProSUM: prospecting secondary raw materials in the Urban mine and mining wastes. Systematic harmonisation and classification of data sources for mapping EU secondary raw materials in electronics, batteries, vehicles and mining wastes. In: Electronics Goes Green Conference, Berlin (2016). ISBN 978-3-00-053763-9
6. Laxton, J., Sen, M., Passmore, J.: The EarthServer geology service: web coverage services for geosciences. In: EGU General Assembly, Vienna, Austria (2014). Id 1543
7. W3C Note: URIs, URLs, and URNs: Clarifications and Recommendations 1.0

Author Index

Printed in the United States
By Bookmasters